Annotated Teacher's Edition

WORLD OF VOCABULARY

AQUA

Sidney J. Rauch

Zacharie J. Clements

Assisted by Barry Schoenholz

World of Vocabulary, Aqua Level, Third Edition

Sidney J. Rauch • Zacharie J. Clements

Copyright © 1996 by Globe Fearon Educational Publisher, a division of Simon & Schuster, One Lake Street, Upper Saddle River, New Jersey 07458. All rights reserved. No part of this book may be reproduced or transmitted in any form or by any means, electrical or mechanical, including photocopying, recording, or by any information storage and retrieval system without permission in writing from the publisher.

Printed in the United States of America

1 2 3 4 5 6 7 8 9 10   99 98 97 96

ISBN: 0-8359-1292-2

CONTENTS

INTRODUCTION

>>>> Program Overview

The eight books in the *World of Vocabulary* series are especially designed to interest ESL/LEP students and students who have been reluctant or slow to expand their vocabularies. As an effective alternative to traditional vocabulary development programs, each lesson in *World of Vocabulary* offers:

- a short, high-interest, nonfiction article that incorporates the key words for that lesson in a meaningful context.
- photographs that hold students' attention and provide additional context for the key words.
- a variety of short skills exercises that build understanding and retention.
- high-interest writing and simple research projects that offer opportunities for students to extend their learning.

This revised edition of *World of Vocabulary* includes new and updated lessons at all levels. Changes in the design make the vocabulary lessons and instructions to students clearer and more approachable. New stories in all eight books spotlight personalities, such as Jim Carrey, Sandra Cisneros, and Steven Spielberg, and cover topics as engaging as World Cup soccer, Navajo code talkers, "Star Trek," and a group of Los Angeles teenagers who make and sell "Food from the 'Hood."

The revised series continues to offer diverse subjects and now includes selections on Latino actor Edward James Olmos, African American writer Walter Dean Myers, Native American ballerina Maria Tallchief, Latino baseball legend Roberto Clemente, Puerto Rican writer Nicholasa Mohr, and Chinese American writer Laurence Yep.

All eight books continue to have color designations rather than numbers to prevent students from identifying the books with grade levels. The use of color levels also enables teachers to provide individual students with the appropriate reading skills and vocabulary enhancement without calling attention to their reading levels. Below are the revised *World of Vocabulary* books listed by color and reading level:

Yellow	3
Tan	4
Aqua	5
Orange	6
Blue	7
Red	8
Purple	9
Green	10

As in the earlier version of *World of Vocabulary*, some lesson elements are carried throughout the series, but the pedagogy and design of each book is geared to the needs of students at that level. For example, the Yellow and Tan books are set in a larger typeface and have more write-on space for student responses than the other six books. Each lesson in these first two books contains eight key vocabulary words, compared to ten key words in the other books.

The Yellow and Tan books also include 15 units rather than the 20 units in the other books in the series. "Using Your Language" exercises in the Yellow and Tan books teach fundamental language skills that may need reinforcement at this reading level. The Tan book adds a phonics exercise.

As the reading level progresses in the next six books, the exercises offer more vocabulary words and increasing challenges. For example, "Find the Analogies" exercises appear in some lessons at the Aqua, Orange, and Red levels but are a part of every lesson in the Blue, Purple, and Green books.

>>>> The Need for Vocabulary Development

Learning depends on the comprehension and use of words. Students who learn new words and add them to their working vocabularies increase their chances for success in all subject areas.

Understanding new words is especially crucial for remedial and second-language learners. Their reluctance or inability to read makes it even more difficult for them to tackle unfamiliar words. The *World of Vocabulary* series was created for these students. The reading level of each selection is carefully controlled so students will not be burdened with an overload of new words.

Most importantly, the *World of Vocabulary* series motivates students by inviting them to relate their own experiences and ideas to the selections. In doing so, students gain essential practice in the interrelated skills of listening, speaking, reading, and writing. This practice and reinforcement enhances their vocabulary and language development.

>>>> Key Strategies Used in the Series

Providing Varied Experiences

The more varied experiences students have, the more meaning they can obtain from the printed word. For example, students who have studied the development of the space program will also have learned many new words, such as *astronaut* and *module*. They have also attached new meanings to old words, such as *shuttle* and *feedback*.

The reading selections in the *World of Vocabulary* series enable students to enrich their vocabulary by exploring major news events, as well as the lives and motivations of fascinating people. Through the wide range of selections, students encounter new words and learn different meanings of old words.

Visual tools are also valuable sources of experience. The full-page and smaller photographs in the lessons capture students' attention and help them to understand the words in the reading selections.

Building Motivation

If we can create the desire to read, we are on our way to successful teaching. Formal research and classroom experience have shown that the great majority of students are motivated to read if the following ingredients are present: opportunities for success, high-interest materials, appropriate reading levels, the chance to work at their own rate, and opportunities to share their experiences.

All of these ingredients are incorporated into the *World of Vocabulary* lessons through the use of engaging reading selections, controlled reading levels, a range of skills exercises, and discussion and enrichment opportunities.

Making Learning Meaningful

We do not often learn new words after one exposure, so vocabulary development requires repetition in meaningful situations. The *World of Vocabulary* series provides opportunities for students to use new words in relevant speaking and writing activities based on the high-interest reading selections.

Fostering Success

When students feel they have accomplished something, they want to continue. The *World of Vocabulary* series is designed to help students gain a feeling of accomplishment through listening, speaking, reading, and writing activities that motivate them to go beyond the lessons.

>>>> Readability Levels

The reading level in each lesson is controlled in two ways. First, vocabulary words appropriate to the designated reading level were selected from the EDL Core Vocabulary Cumulative List. The words were chosen for their interest, motivational level, and relevance to each reading selection.

Next, the reading level of each selection was adjusted using the Flesch-Kincaid Readability Index. This formula takes into account average sentence length, number of words per sentence, and number of syllables per word.

VOCABULARY STRATEGIES

>>>> Learning and Thinking Styles

People of all ages learn and think in different ways. For example, most of us receive information through our five senses, but each of us tends to prefer learning through one sense, such as our visual or auditory modality.

By keeping in mind the different ways students learn and think, we can appeal to the range of learning and thinking styles. By taking different styles into account in planning lessons, we can help all students understand new information and ideas and apply this knowledge and insight to their lives.

There are three main learning styles:
- Visual learners like to see ideas.
- Auditory learners prefer to hear information.
- Kinesthetic or tactile learners absorb concepts better when they can move about and use their hands to feel and manipulate objects.

After we receive information, we tend to process or think about the information in one of two ways:
- Global thinkers prefer to see the "big picture," the whole idea or the general pattern, before they think about the details. They search for relationships among ideas and like to make generalizations. They are especially interested in material that relates to their own lives. Global thinkers tend to be impulsive and quick to respond to teachers' questions.
- Analytical thinkers focus first on the parts and then put them together to form a whole. They think in a step-by-step approach and look at information in a more impersonal way. They are more likely to analyze information and ideas rather than apply it to their own lives. Analytical thinkers tend to be reflective and thoughtful in their answers.

However, few of us are *only* auditory learners or *only* analytical thinkers. Most people use a combination of learning and thinking styles but prefer one modality or style over the others. An effective lesson takes into account all three types of learning and both types of thinking. The ideas below, in addition to your own creativity, will help you meet the needs and preferences of every student in your class.

Visual Learners
- Write the lesson's key vocabulary words on the chalkboard, overhead transparency, or poster so students can see the words and refer to them.
- Encourage students to examine the photographs in the lesson and to explain what the pictures tell them about the key words.
- Repeat oral instructions or write them on the board. After giving instructions, put examples on the board.

- Involve students in creating word cluster maps (see p. xiv) to help them analyze word meanings.
- Use the other graphic organizers on pp. xiii-xvii to help students put analogies and other ideas into a visual form.
- Display some of the writing assignments students complete for the "Learn More About..." sections. Encourage students to read each other's work.
- For selections that focus on authors or artists, collect books, pictures, or other works by that author or artist for students to examine.
- For selections that focus on actors, show videotapes of their movies or television shows.

Auditory Learners

- Invite a volunteer to read aloud the selection at the beginning of each unit as students follow along in their books. You might audiotape the selection so students can listen to it again on their own.
- Ask a student to read aloud the "Understanding the Story" questions.
- Provide time for class and small-group discussions.
- Read aloud the directions printed in the books.
- Occasionally do an activity orally as a class, such as "Complete the Story."
- Allow students to make oral presentations or to audiotape assignments from the "Learn More About..." sections.

Kinesthetic or Tactile Learners

- Encourage students to take notes so the movements of their hands can help them learn new information.
- Encourage students to draw pictures to illustrate new words.
- In small groups, have students act out new words as they say the words aloud.
- Invite students to clap out the syllable patterns and/or spellings of new words.

- Write (or have students write) the new words on cards that can be handled and distributed.
- Provide sets of letters that students can arrange to spell the key words.

Global Thinkers

- Explain the "big picture," or the general idea, first.
- Point out how the key words fit patterns students have studied and how they relate to words and concepts that are already familiar to students.
- Involve students in brainstorming and discussion groups. Encourage students to express ideas and images that they associate with the new words.
- Explore ways that ideas and information are relevant to students.
- Encourage students to think about their answers before they respond.
- Set goals and offer reinforcement for meeting those goals.

Analytical Thinkers

- Start with the facts and then offer an overview of the topic.
- Give students time to think about their answers before they respond.
- Encourage students to set their own goals and to provide their own reinforcement for meeting them.
- Suggest that students classify new words into several different categories.
- Provide time for students to organize concepts or processes in a step-by-step approach.
- Help students recognize how new concepts relate to their own lives.

>>>> Cooperative Learning

One way to address multiple learning and thinking styles and to engage students more actively in their own learning is through cooperative learning activities.

Cooperative learning means more than having students work in groups. They must work together toward a shared goal that depends on each person's contribution. In cooperative learning, group members share ideas and materials, divide task responsibilities among themselves, rely on each other to complete these responsibilities, and are rewarded as a group for successful completion of a task.

If your students are not accustomed to group work, you might assign (or have students choose) group roles, such as discussion leader, recorder, reporter, or timekeeper. Having specific responsibilities will help group members work together.

Cooperative learning has many applications in the *World of Vocabulary* series. For instance, you might organize the class into groups and have each group teach its own members the key vocabulary words in that lesson. Groups could use a jigsaw approach, with each person learning and then teaching two or three words to other members of the group. Groups might create their own word searches, flashcards, crossword puzzles, incomplete sentences, analogies, and so on.

Then evaluate each group member to determine his or her level of understanding. Or you might ask group members to number off so you can evaluate only the 3s, for example. Explain that you will hold the entire group accountable for those students' mastery of the lesson words.

In other applications of cooperative learning, students might work together to create one product, such as a cluster map, a simple research project, or an original story that incorporates the key vocabulary words.

You might also consider trying the cooperative learning activities below, modifying them so they will be appropriate for your students.

Word Round-Robin

Organize the class into groups of ten (eight for Yellow and Tan levels) and have each group sit in a circle. Ask members to count off 1-10 (or 1-8) and give everyone a sheet of paper. Assign all the 1s one vocabulary word from the lesson, the 2s another word, and so on. Then follow these steps:

Step 1: Ask students to write their assigned word and their best guess as to its definition.

Step 2: Have students pass their papers to the person on their right. Then tell them to read through the story to find the word on the paper they received. Have students write another definition below the first definition on the paper, using context clues from the story.

Step 3: Ask students to pass their papers to their right. This time tell students to use a dictionary to look up the word on the paper they received. Then have them write on the paper the dictionary definition and a sentence that includes the word, using the same meaning as in the story.

Step 4: Invite groups to read each paper aloud, discuss the word, and write one definition in their own words, based on what members wrote on the papers.

Step 5: Have each group share its definition for the assigned word with the class. Discuss similarities and differences among the definitions. Guide students to recognize that definitions of some new words are clear even in isolation because of their root words, while others have multiple definitions that depend on the context in which they are used.

Synonym Seekers

Involve the class in preparing for this activity by assigning a vocabulary word to each pair of students. (You might include words from more than one selection.) Each pair will write its word and as many synonyms as possible on an index card, consulting a dictionary and thesaurus, if you wish.

Have pairs share their cards with the class, explaining subtle differences among the synonyms. Then collect the cards and combine pairs of students to form teams of four or six. Call out one of the vocabulary words and ask teams to write down as many synonyms as they can think of in 30 seconds.

Then read the synonyms listed on the card. Teams will give themselves one point for each synonym they recalled. Encourage students to suggest new synonyms to add to the cards and discuss why certain words could not be used as synonyms. Play the game several times with these cards before creating new cards with other words.

>>>> Approaches for ESL/LEP Students

- Invite volunteers to read the stories aloud while students follow along in their books.
- Watch for figurative expressions in the lessons and discuss their literal and intended meanings. Examples include "making faces," "friendly fire," and "bounce off the walls."
- Help students identify root words. Involve them in listing other words with the same roots and in exploring their meanings.
- Compare how prefixes and suffixes in the vocabulary words are similar to those in words students already know.
- Make word webs to help students understand relationships among words and concepts. Use the graphic organizer on page xiv or write a vocabulary word in the center of the chalkboard or poster. Invite students to name as many related words as possible for you to write around the key word. Discuss how each word is related.
- Involve students in listing words that are similar in some way to a vocabulary word, such as other vehicles, adverbs, occupations, and so on.
- Encourage students to share words or phrases from their native languages that mean the same as the vocabulary words. Invite them to teach these words from their native languages to the class.
- Arrange cooperative learning and other activities so ESL/LEP students are grouped with students who speak fluent English.
- Periodically group ESL/LEP students together so that they can assist one another in their native languages.
- Foster discussion with questions, such as "Do you think our space program should send more astronauts to the moon? Why?" and "Would you like to perform in a circus? Why?" These kinds of questions encourage students to use English to share their ideas and opinions.

>>>> Cross-Curricular Connections

General
- Challenge students to identify vocabulary words that have different meanings in other subject areas. For example, *fins* are defined as "rubber flippers" in the Aqua book. How are *fins* defined in science?
- Give extra credit to students who find the lesson's vocabulary words in other textbooks or in newspapers and magazines. Discuss whether the meaning is the same in both uses.

Math
- Invite pairs of students to write problems that include vocabulary words. The difficulty level will depend on their math skills. Ask pairs to exchange problems and try to solve each others'.

Language Arts
- Encourage students to write letters to some of the people described in the stories. Ask them to incorporate some of the lesson's key words into their letters.
- Have students, working in pairs or individually, write their own stories, using a certain number of vocabulary words from one or more selections.

They might leave the spaces blank and challenge other students to complete the stories correctly.

- Organize a spelling contest, using vocabulary words.
- Have groups prepare crossword puzzles that will be combined into a book.
- Encourage students to conduct surveys and/or interview people regarding topics that stem from the stories. For example, how many students or staff at school collect trading cards? What kinds do they collect? How many students or staff are "Star Trek" fans? What attracts them to "Star Trek"? Encourage students to graph their findings and to write short reports explaining their conclusions.

A SAMPLE LESSON PLAN

The following is a suggested plan for teaching a lesson from the *World of Vocabulary* series. You might use it as a guide for preparing and organizing your lessons. However, be sure to modify it where necessary to take into account your students' needs, abilities, interests, and learning styles, along with the specific exercises included in that lesson.

>>>> Setting Objectives

Each lesson in the *World of Vocabulary* series is based on the objectives below.

- To create enthusiasm for and an understanding of the importance of learning new words
- To improve reading comprehension by teaching the meanings of new words, stressing context clues
- To improve vocabulary by presenting key words in exercises that range from simple to complex and that allow for reinforcement of learning
- To encourage oral expression and motivate further study by introducing a highly interesting topic.

>>>> Stimulating Interest

Invite students to examine the photograph on the first page of the lesson. To stimulate their curiosity and involve them in the topic, ask questions. For example, if the lesson were about Koko the gorilla, you might ask:

- The gorilla in the picture is named Koko. How do you think Koko might be different from other gorillas?
- Do you think it is possible to teach a gorilla to talk? Why or why not?
- If Koko could talk to people, what do you think she might say?

>>>> Reading the Story

Have students read the story, silently or in small groups. You also might assign the story to be read outside class. To help auditory learners and ESL/LEP students, ask a volunteer to read the story aloud while classmates follow along in their books.

Encourage students to use the context clues in the story and the opening photograph to figure out the meanings of several boldfaced words. You might have students suggest a definition for each key word, based on context clues. Write the definitions on the chalkboard so the class can review and modify them later in the lesson.

As an aid to ESL/LEP students, discuss words or phrases in the story that have more than one meaning or that have figurative meanings. Two examples in the story about Koko are "blew kisses" and "spends some time."

>>>> Completing the Exercises

The information about exercises below is based on the lesson about Koko in the Orange level. However, books at different levels include different exercises. For example, the Yellow and Tan books offer a simpler activity called "Make a List" instead of the "Understanding the Story" exercise.

Students using the Yellow and Tan books also complete an exercise called "Find the Synonyms," while students at the Aqua, Orange, and Red levels have the "Complete the Sentences" exercise. The equivalent exercise for students at the Blue, Purple, and Green levels is called "Find the Analogies." Each level also includes a variety of other grammar and skills exercises.

Despite variations in exercises from level to level, the explanations below will help you understand why certain exercises are included and how they can be modified to support different learning and thinking styles.

>>>> "Understanding the Story"

This exercise usually asks students to determine the main idea of the selection and to make an inference as a way of assessing their general understanding of the story. Remember that global thinkers may have an easier time describing the main idea than analytical thinkers, who tend to focus on the parts of the story rather than the whole idea.

To use this as a cooperative activity, have students discuss the questions in groups of two or three. Then pair two groups so they can share their conclusions. Ask groups that disagreed on the answers to tell the class the reasoning for their different choices. Be sure to clear up any misunderstandings that become apparent without squelching creativity.

To make sure all students understand the general content of the story, ask a volunteer to summarize it in a sentence or two. Then give the analytical thinkers in the class an opportunity to contribute by describing some of the key supporting details in the story.

>>>> "Make an Alphabetical List"

This activity encourages students to study the key words closely and to become more familiar with their spellings. Practicing writing the words in alphabetical order will be especially beneficial for kinesthetic learners.

To check students' accuracy in arranging the words in alphabetical order, ask one or two students to read their lists aloud. Visual learners will appreciate seeing the list written on the board.

If necessary, model the pronunciation of certain words. Practice saying the more difficult words as a class. (This technique will also be helpful for ESL/LEP students.)

>>>> "What Do the Words Mean?"

In this exercise, students match the definitions listed in their books to the lesson's vocabulary words. If students offer other definitions for the same words, encourage them to consult a dictionary to check their accuracy. Many of the key words have different meanings in other contexts.

Encourage students to suggest synonyms for the words and perhaps some antonyms. Analytical learners might enjoy identifying root words and listing other words with the same roots, prefixes, or suffixes. Many ESL/LEP students will also benefit from this analysis.

>>>> "Complete the Sentences"

This exercise gives students another opportunity to practice using context clues as they complete a set of sentences using the key words.

>>>> "Use Your Own Words"

Working individually or in groups, students are encouraged to brainstorm words that describe a picture or express their reactions to it. This exercise fosters creativity involves students in the lesson by asking for their personal responses. Their responses will depend on their prior knowledge and individual perceptions, so answers are not included in the Answer Key. You might use some of the graphic organizers on pages xiii–xvii for this activity.

As a cooperative activity, students might enjoy working with three classmates to write a group description of the picture. Tell the first group member to write a word related to the picture on a sheet of paper and to pass the paper to the right. Have the next two group members add their own words, different from the ones already listed. Then ask the fourth group member to write one sentence about the picture that includes all three words. Start another sheet of paper with a different group member and continue in the same way, with the fourth member combining the words into one sentence.

>>>> "Make New Words from Old"

This is one of several reinforcement exercises throughout the *World of Vocabulary* series. "Make New Words from Old" invites students to look creatively at the letters in a key word from the lesson. Other exercises in the series challenge students to identify synonyms and/or antonyms, underline specific parts of speech, to find the subjects and predicates in sentences, to write the possessive forms of words, or to complete other activities that focus on key words from the lesson.

>>>> "Complete the Story"

Students again use context clues to place the lesson's key words correctly in a new story. This story relates to the one that opened the lesson and may offer more information on the topic or encourage students to apply new knowledge or insights in their own lives. You might use "Complete the Story" as a post-test of student mastery of the key words.

>>>> "Learn More About..."

The last page of each lesson offers one to four activities that encourage students to learn more about the lesson's topic. You might assign one or several activities or encourage students to choose an activity to complete for extra credit. They could work during class time or outside of class—individually, with partners, or in small groups.

Some of the activities are developed for ESL/LEP students, while others provide opportunities for cooperative learning, cross-curricular projects, and enrichment. Placing activities in these categories was not meant to limit their use, as many of the activities would benefit and interest most students. For some reluctant readers, these projects may be their first attempt at independent research, fueled by their interest in the lesson's topic.

Some lessons include a "Further Reading" activity that lists fiction or nonfiction books on the lesson's topic that are appropriate for that reading level. Students are asked to complete a brief activity after their reading.

"Further Reading" and other activities that require a written response provide additional opportunities for students to practice and receive feedback on their writing skills, including punctuation, capitalization, and spelling. The effort students spend on the "Learn More About" activities can result in marked improvements in their reading and writing skills.

>>>> The topic of the story:

>>>> The main idea of the story:

>>>> Some details from the story:

>>>> What interested me most:

>>>> A question I would like to ask:

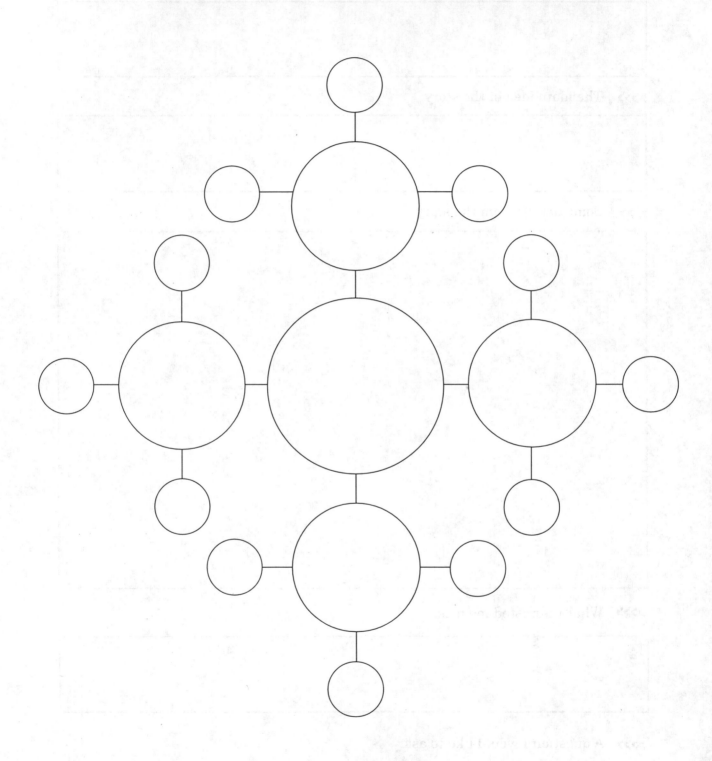

A _____ is a _____ that
   (key word)   (description)

_____, _____, and
 (characteristic)   (characteristic)

_____.
 (characteristic)

ANALOGY ORGANIZER

relationship

_____ is to _____ [ ] as _____ is to _____.

relationship

_____ is to _____ [ ] as _____ is to _____.

relationship

_____ is to _____ [ ] as _____ is to _____.

relationship

_____ is to _____ [ ] as _____ is to _____.

relationship

_____ is to _____ [ ] as _____ is to _____.

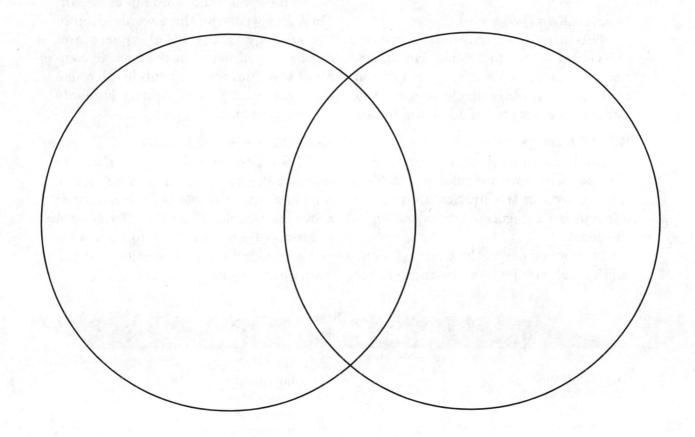

Understanding the Story (page xiii)

This form will help students analyze the selection at the beginning of each unit and organize its content logically and visually.

Cluster Map (page xiv)

This organizer will help structure individual or group brainstorming sessions. Students write a key word or concept in the middle circle and related words or concepts in the adjoining circles.

Word Chart (page xv)

Students can use this chart to compare key words with similar words. They write words in the first column on the left and list categories across the top of the chart.

In another use of the chart, students might write adjectives in the left column. Across the top, they could list experiences.

Analogy Organizer (page xvi)

Students write in the box the relationship between the words in each pair. Then they write the three words given in the analogy in the blank spaces and choose a word listed in the book to complete the analogy. You might also use this graphic organizer to help students write their own analogies.

Venn Diagram (page xvii)

This organizer offers another way to compare the meanings of similar words. Students write the words to be compared above each circle. Then they list several properties that apply only to each word inside the circle and properties that the words share in the center area.

Aqua Test Answers

1. blizzards
2. shadows
3. rarely
4. bruises
5. created
6. constantly
7. competitors
8. tasks
9. bold
10. glow
11. inflated
12. bred
13. preparation
14. breathe
15. scheduled
16. spread
17. celebration
18. species
19. movement
20. tickets
21. advantage
22. different
23. plan
24. earnings
25. decreased
26. enjoyment
27. fashion
28. attention
29. express
30. cruel
31. honest
32. uncertain
33. skills
34. discuss
35. purchase
36. professional

AQUA TEST

WHAT DO THE WORDS MEAN?

>>>> *Following are some meanings, or definitions, for the ten vocabulary words in the box below. Write the words next to their definitions.*

competitors	created	constantly	bold	glow
bruises	shadows	rarely	tasks	blizzards

1. _____ extreme and violent snowstorms

2. _____ areas of darkness or shade

3. _____ not often; seldom

4. _____ marks caused by an injury that does not break the skin

5. _____ invented

6. _____ repeatedly; happening again and again

7. _____ contestants

8. _____ jobs; assignments

9. _____ not afraid of danger; brave

10. _____ a light; to shine

COMPLETE THE SENTENCES

>>>> *Use the vocabulary words in this lesson to complete the following sentences. Use each word only once.*

breathe	preparation	bred	species	tickets
inflated	scheduled	spread	movement	celebration

11. At the gas station, we got our tires _____.

12. These roses are _____ to smell good.

13. A camping trip requires a lot of _____.

14. The smog was so thick I could hardly _____.

15. When is the girls' swim meet _____?

16. My poison ivy has _____ to my face!

17. We are planning a _____ for my sister's sixteenth birthday.

18. Dogs and cats are from two different _____.

19. I have joined the _____ to save rare animals.

20. My brother already has two speeding _____.

CIRCLE THE SYNONYMS

>>>> A **synonym** is a word that means the same or nearly the same as another word. Below are six vocabulary words. They are each followed by four other words or groups of words. One of these words is a synonym; the others are not. Draw a circle around the word that is a synonym.

Vocabulary Words	Synonyms			
21. **benefit**	admiration	brilliant	advantage	included
22. **various**	large	different	decreasing	foreign
23. **system**	plan	stereo	depth	obstacles
24. **income**	movement	division	earnings	design
25. **reduced**	charmed	decreased	detailed	interested
26. **pleasure**	attention	energy	purposes	enjoyment

COMPLETE THE STORY

>>>> *Use words from the box to fill in the blanks and complete the story. Use each word only once.*

express	cruel	honest	discuss	professional
attention	fashion	uncertain	skills	purchase

Sometimes I have the feeling that my clothes will never be in (**27**) _____ . I guess I need to pay more (**28**) _____ to the current styles. I try to wear clothes that (**29**) _____ my personality, but sometimes my friends tease me about them. They do not mean to be (**30**) _____, but I feel embarrassed anyway.

To be (**31**) _____, I feel (**32**) _____ when I shop for clothes. I just never developed (**33**) _____ in choosing clothes. I need someone to (**34**) _____ the choices with me and help me (**35**) _____ the right ones. If there are any (**36**) _____ shoppers out there, give me a call!

XX

CONTENTS

1 THIS JOKER IS WILD!

Robin Williams went to college to study economics. He took a class in which the students had to *improvise* speeches. He enjoyed using his imagination and speaking without notes. He felt he had discovered the thing he "was meant to do."

Williams dropped out of college. He became a stand-up comic. This led to a small role on the television show "Happy Days." People loved the wild character he played. Soon, Williams had a show of his own called "Mork and Mindy."

After conquering television, Williams moved his *career* to films. His first movies were *disappointing.* Williams's magic didn't seem to work on film. He was under control and seemed *clumsy.* Then Williams agreed to do *Good Morning, Vietnam*. This film's writers used a trick from "Mork and Mindy." They allowed Williams to improvise his lines. That was all Williams needed. He *created* original comic routines as he went along! The result was a successful movie.

Williams went on to star in such hit movies as *Dead Poet's Society, Awakenings, The Fisher King, Hook*, and *Mrs. Doubtfire*. With each role, he has grown as an actor. In his early movies, he was *nervous* in front of the movie cameras. Then he learned how to use the cameras to his *benefit.* One of Williams's most popular roles is one in which he is never seen on *screen.* He is the voice of the genie in the animated movie *Aladdin*. As with "Mork and Mindy," he improvised many of his lines. His *energy* helped to make the movie a big hit. Today, Williams is *respected* for his comic ability.

UNDERSTANDING THE STORY

 Circle the letter next to each correct statement.

1. Robin Williams is known for his
   a. ability to improvise.
   b. constant comedy writing.
   c. early film successes.
2. Audiences probably enjoy Williams because
   a. he is never sad.
   b. they don't know what he might do next.
   c. he makes fun of them.

3

MAKE AN ALPHABETICAL LIST

>>>> Here are the ten vocabulary words in the lesson. Write them in alphabetical order in the spaces below.

career	nervous	screen	respected	disappointing
improvise	clumsy	energy	created	benefit

1. benefit
2. career
3. clumsy
4. created
5. disappointing

6. energy
7. improvise
8. nervous
9. respected
10. screen

WHAT DO THE WORDS MEAN?

>>>> Following are some meanings, or definitions, for the ten vocabulary words in this lesson. Write the words next to their definitions.

1. nervous — uneasy; uncomfortable
2. screen — a surface or area on which movies or television images are shown
3. created — invented
4. respected — admired
5. clumsy — not graceful
6. disappointing — not satisfying
7. career — occupation; work
8. improvise — to make up
9. energy — enthusiasm; an inner power or ability
10. benefit — advantage

4

COMPLETE THE SENTENCES

>>>> *Use the vocabulary words in this lesson to complete the following sentences. Use each word only once.*

career	nervous	screen	respected	disappointing
improvise	clumsy	energy	created	benefit

1. Williams seems relaxed today, but cameras still make him _____nervous_____.

2. Williams's _____career_____ includes live appearances, television, and films.

3. Williams's first films were _____disappointing_____ to his fans.

4. In his first films, Williams was _____clumsy_____ in front of the camera.

5. Mork was _____created_____ by scriptwriters on "Happy Days."

6. Williams never actually appears on _____screen_____ in the movie *Aladdin*.

7. Williams likes to _____improvise_____ his own comedy routines in comedy clubs.

8. He has learned to use movie cameras to his own _____benefit_____.

9. He has so much _____energy_____ that he seems wild and out of control.

10. Williams is _____respected_____ because he works so hard.

USE YOUR OWN WORDS

>>>> *Look at the picture. What words come into your mind other than the ten vocabulary words used in this lesson? Write them on the lines below. To help you get started, here are two good words:*

1. _____headphones_____
2. _____talking_____
3. _____Answers will vary._____
4. _____
5. _____
6. _____
7. _____
8. _____
9. _____
10. _____

IDENTIFY THE SYNONYMS AND ANTONYMS

>>>> There are six vocabulary words listed below. To the right of each is either a synonym or an antonym. Remember: a **synonym** is a word that means the same or nearly the same as another word. An **antonym** is a word that means the opposite of another word.

>>>> *On the line beside each pair of words, write **S** for synonyms or **A** for antonyms.*

1. **disappointing**	pleasing	1.	A
2. **clumsy**	awkward	2.	S
3. **created**	destroyed	3.	A
4. **nervous**	calm	4.	A
5. **respected**	admired	5.	S
6. **benefit**	advantage	6.	S

COMPLETE THE STORY

>>>> Here are the ten vocabulary words for this lesson:

career	nervous	screen	respected	disappointing
improvise	clumsy	energy	created	benefit

>>>> *There are five blank spaces in the story below. Five vocabulary words have already been used in the story. They are underlined. Use the other five words to fill in the blanks.*

Robin Williams is always entertaining. He is never _____disappointing_____. He likes to improvise comedy routines on every subject. Williams seems to love working live. The eager audience doesn't make him nervous. In fact, he seems to draw his _____energy_____ from audience support. He gets wilder and wilder as they cheer. His work seems to benefit from their encouragement.

Williams loves working live. He still likes to surprise audiences. However, he has been in both television and films in his _____career_____, too. His work is respected, but it was not always a success. At first, people were not sure about his films. The camera made Williams seem _____clumsy_____. He seemed unsure about what to do. Most people preferred his live comedy. They did not like the serious characters he played on the _____screen_____. He seems unworried. He has created a secure place for his work.

Learn More About Being Funny

>>>> *On a separate piece of paper or in your notebook or journal, complete one or more of the activities below.*

Building Language

Think of three jokes you might know in another language. Now, translate them into English. Tell them to a friend in English. Did the friend think they were funny? Write whether the jokes were funny in English. If your friend did not think they were funny, try to explain why the joke did not translate well into English.

Broadening Your Understanding

Robin Williams is known as a brilliant stand-up comic. In a stand-up routine, a comedian strings together a series of jokes. Often, the jokes have to do with things that happen to people in real life. Write your own three-minute stand-up comedy routine. Then try it out on a friend.

Extending Your Reading

Why would someone want to be a comedian? Read one of the following books to find out what causes a person to try for a career in comedy. Then write about why you would or wouldn't want a career as a comedian.

Will Rogers, by Liz Sonneborn
Bill Cosby, by George H. Hill
Eddie Murphy, by Deborah Wilburn
Whoopi Goldberg, by Mary Agnes Adams

2 THE OLDEST FLYING MACHINE

It is cool and quiet. The ground is far below. You are slowly floating through the air. You are in a giant **helium** balloon. Balloons are the oldest form of flying machine.

In early times, balloons were made of silk. These balloons were **inflated** with hot air. To make the air hot, a small fire was built in a special stove in the **gondola.** The gondola, or basket, was hung from the balloon by ropes. People rode in the gondola.

Ballasts, or weights made of sand-filled bags, held the balloon down. When enough ballasts were taken off, the balloon rose in the air. Pilots had to avoid **ascending** too rapidly because the balloon could **rupture.** This rapid ascent would cause all the hot, light air to escape. The gondola would crash to the ground. When the balloon was **aloft,** fire was a constant **threat.**

How about coming down? Simple—the fire was put out. As the air in the balloon cooled, the balloon came down. The pilot had to **descend** slowly. If the balloon came down too fast, the gondola could be smashed. Early balloonists were both **bold** and skillful.

UNDERSTANDING THE STORY

>>>> *Circle the letter next to each correct statement.*

1. The main purpose of this story is to describe
   a. the different kinds of balloons available.
   b. how a hot-air balloon works.
   c. the very first flight made in a balloon.

2. Though it doesn't say so, from the story you get the idea that
   a. there were many accidents when balloons were first used.
   b. a person needed a license to fly a balloon.
   c. balloons were first flown in the United States.

MAKE AN ALPHABETICAL LIST

>>>> *Here are the ten vocabulary words in the lesson. Write them in alphabetical order in the spaces below.*

helium	inflated	gondola	ballasts	ascending
rupture	aloft	threat	bold	descend

1. aloft
2. ascending
3. ballasts
4. bold
5. descend

6. gondola
7. helium
8. inflated
9. rupture
10. threat

WHAT DO THE WORDS MEAN?

>>>> *Following are some meanings, or definitions, for the ten vocabulary words in this lesson. Write the words next to their definitions.*

1. helium — gas used to fill balloons
2. inflated — filled up
3. gondola — a car or basket hung under a balloon
4. ballasts — weights used to make a gondola heavier
5. ascending — going up
6. rupture — to break open; to burst
7. aloft — high; above the earth
8. threat — something dangerous that might happen
9. descend — to go down
10. bold — not afraid of danger; brave

COMPLETE THE SENTENCES

>>>> *Use the vocabulary words in this lesson to complete the following sentences. Use each word only once.*

ballasts	helium	descend	inflated	bold
rupture	ascending	threat	gondola	aloft

1. Because _____helium_____ is lighter than air, it keeps the balloon aloft.

2. The excited riders in the _____gondola_____ waved at the people below.

3. The danger in _____ascending_____ too rapidly is that the balloon might burst.

4. A person had to be _____bold_____ and skillful to take a balloon out in the early days because ballooning was so dangerous.

5. Sand-filled bags called _____ballasts_____ are released to allow the balloon to rise.

6. A balloon must _____descend_____ slowly if the landing is going to be smooth.

7. Fire was a constant _____threat_____ to the safety of the early balloonists.

8. When you are _____aloft_____ in a balloon, you can see towns many miles away.

9. Children used to gather to watch the balloons being _____inflated_____ before a trip.

10. The pilot became worried when she noticed a small _____rupture_____ in the balloon.

USE YOUR OWN WORDS

>>>> *Look at the picture. What words come into your mind other than the ten vocabulary words used in this lesson? Write them on the lines below. To help you get started, here are two good words:*

1. _____sky_____
2. _____ground_____
3. _____Answers will vary._____
4. _____
5. _____
6. _____
7. _____
8. _____
9. _____
10. _____

UNSCRAMBLE THE LETTERS

>>>> *Each group of letters contains the letters in one of the vocabulary words for this lesson. Can you unscramble them? Write your answers in the lines to the right of each letter group.*

Scrambled Words	Vocabulary Words
1. ulihem	helium
2. gsadicnen	ascending
3. folta	aloft
4. laltasbs	ballasts
5. eruturp	rupture
6. dalogon	gondola
7. lafindet	inflated
8. dolb	bold
9. rathet	threat
10. secdned	descend

COMPLETE THE STORY

>>>> Here are the ten vocabulary words for this lesson:

helium	ballasts	gondola	aloft	inflated
threat	bold	ascending	descent	rupture

>>>> *There are five blank spaces in the story below. Five vocabulary words have already been used in the story. They are underlined. Use the other five words to fill in the blanks.*

Going up in a balloon is great fun for some people. You step into the ___gondola___. The <u>ballasts</u> are dropped, and you begin ___aloft___. Once ___ascending___, you can see the ground far below. In the old days, you had to be a <u>bold</u> person to try ballooning. There was always the <u>threat</u> of fire. Care had to be taken not to <u>rupture</u> the delicate balloon skin. When you were ready to ___descend___, you would put out the fire. Today balloons are <u>inflated</u> with ___helium___.

Learn More About Flying Machines

>>>> *On a separate piece of paper or in your notebook or journal, complete one or more of the activities below.*

Learning Across the Curriculum

Read about the science behind the flying machines that people use today. Write an explanation of how a hot air balloon, a plane, or a helicopter flies. You may want to use diagrams in your report.

Broadening Your Understanding

The Albuquerque International Balloon Fiesta in October is the biggest gathering of hot-air balloons in the United States. Every year, more than 500 hot-air balloons and their owners participate. Another 1.5 million people come to the nine-day program. Write to the Balloon Fiesta at 8309 Washington Place NE, Albuquerque, New Mexico 87113, and ask for information about the festival. Then plan a trip there, using guidebooks if you need to. Include where you will stay, how long you will stay, what you will do every day, and how much the trip will cost.

Extending Your Reading

Use one of the following books to help you make a kite. After you make your kite, be sure it flies. Then write the scientific reasons why your kite stays in the air.

Making Kites, by David Michael
Dynamite Kites, by Jack Wiley and Suzanne L. Cheatle
Better Kite Flying, by Ross Olney
Kites for Kids, by Burton and Rita Marks

Do you picture *senior* citizens sitting in rocking chairs and sharing memories? Well, hold on to that rocker because the *National* Senior Olympics will change your mind. The first senior games were held about 20 years ago in Los Angeles, California, but in 1987, the United States National Senior Olympics organized a *biennial* event. *Competitors* must be more than 55 years old and *qualify* by competing in local *contests.*

In 1989, there were about 3,500 athletes from 47 states, Puerto Rico, and three *foreign* countries. They participated in 14 sports, from archery to volleyball. The oldest man was 91, and the oldest woman was 87.

One competitor got an early start. He rode his bicycle more than 1,000 miles to reach the games in St. Louis, Missouri. Of course, he was only 64! Foot races seem to bring out the best in these athletes. One racer, who is 79 and blind, says, "I love the excitement and the people." Another competitor is a 64-year-old nun. She raced in her *habit,* with her rosary beads swinging in time to her *stride.* She took up ice skating at 52 and plans to practice the *javelin* throw next because she discovered at a recent meet that she has a natural talent for it. Forget the rocking chair. You'd better keep in shape if you want to keep up with these golden agers!

UNDERSTANDING THE STORY

 Circle the letter next to each correct statement.

1. The main idea of this story is that
   a. Olympic games are very challenging.
   b. senior citizens can lead very active lives.
   c. people like to travel to compete in the games.

2. The fact that 3,500 people participated in the Senior Olympics tells you that
   a. younger athletes should try to help these senior citizens.
   b. we need to build many new sports stadiums.
   c. many senior citizens are interested in keeping in shape.

MAKE AN ALPHABETICAL LIST

>>>> *Here are the ten vocabulary words in the lesson. Write them in alphabetical order in the spaces below.*

senior	contests	national	foreign	biennial
habit	competitors	stride	qualify	javelin

1. biennial
2. competitors
3. contests
4. foreign
5. habit

6. javelin
7. national
8. qualify
9. senior
10. stride

WHAT DO THE WORDS MEAN?

>>>> *Following are some meanings, or definitions, for the ten vocabulary words in this lesson. Write the words next to their definitions.*

1. national — having to do with a whole country

2. foreign — coming from another country

3. senior — older; more than age 55

4. contests — organized sports events

5. habit — an outfit worn by a nun

6. biennial — happening every two years

7. stride — a step or style of walking

8. qualify — to prove oneself worthy

9. javelin — a spear used in sports events

10. competitors — people who play in a contest or game

16

COMPLETE THE SENTENCES

>>>> *Use the vocabulary words in this lesson to complete the following sentences. Use each word only once.*

senior	contests	national	foreign	biennial
habit	competitors	stride	qualify	javelin

1. The original Olympics were athletic ___contests___ in Greece thousands of years ago.

2. These were ___national___ games in which only Greeks could compete.

3. The ___competitors___ were the finest athletes from each city in Greece.

4. Later, there were some ___foreign___ contestants, but most were Greek.

5. An athlete trained carefully to ___qualify___ for these Olympic games.

6. Contestants threw the ___javelin___ and boxed, among other events.

7. A special ___stride___ helped runners complete the long and difficult marathon.

8. The original Olympics were not ___biennial___ events; they were held every four years, as modern Olympics are.

9. Of course, ___senior___ citizens did not participate in these athletic games.

10. On the other hand, Greek athletes didn't have to compete wearing a nun's ___habit___.

USE YOUR OWN WORDS

>>>> *Look at the picture. What words come into your mind other than the ten vocabulary words used in this lesson? Write them on the lines below. To help you get started, here are two good words:*

1. ___excited___
2. ___healthy___
3. ___Answers will vary.___
4. _____
5. _____
6. _____
7. _____
8. _____
9. _____
10. _____

FIND THE ANTONYMS

>>>> **Antonyms** are words that are opposite in meaning. For example, *good* and *bad* are antonyms.

>>>> **Here are antonyms for six vocabulary words. See if you can identify the vocabulary words and write them in the spaces on the left.**

	Vocabulary Words	**Antonyms**
1.	foreign	native
2.	qualify	fail
3.	senior	young
4.	national	local
5.	contests	partnerships
6.	competitors	helpers

COMPLETE THE STORY

>>>> Here are the ten vocabulary words for this lesson:

senior	contests	national	foreign	biennial
competitors	stride	habit	qualify	javelin

>>>> **There are five blanks in the story below. Five vocabulary words have already been used in the story. They are underlined. Use the other five words to fill in the blanks.**

How do _____ senior _____ citizens know that older people like themselves can take part in these Olympics? Well, every two years, there is <u>biennial</u> publicity. Also, local _____ contests _____ encourage older people to become <u>competitors</u>. To go to the _____ national _____ games, such as those in St. Louis, athletes have to _____ qualify _____ by winning local events. Of course, there are always some <u>foreign</u> athletes from other countries.

You don't have to do it the hard way. You can wear warm-up suits instead of a nun's <u>habit</u>. You can practice your <u>stride</u> for races a little bit at a time. You can learn how to hold the _____ javelin _____ before you try to throw it. The important thing to work on is the spirit of friendly competition at these games.

18

Learn More About Older Americans

>>>> *On a separate piece of paper or in your notebook or journal, complete one or more of the activities below.*

Learning Across the Curriculum

Scientists are making discoveries every day about how we can live longer and healthier lives. Read about some of these discoveries. Write ten guidelines that anyone can follow to help them live a longer and healthier life.

Broadening Your Understanding

Imagine you have to find sponsors for the Senior Olympics. Think about what company would benefit by supporting the Senior Olympics. Then write a letter designed to interest a likely sponsor. Explain what the program is and why getting involved with it will help the sponsor, as well as the Senior Olympics.

Extending Your Reading

The following books contain biographies of people who have succeeded. Read one of them and report on what three or four of these people accomplished when they were older.

Great Lives, by William Jay Jacobs
Women of Courage, by Dorothy Nathan
People Who Make a Difference, by Brent Ashabranner

There is a city in Europe that does not have a paved street. It does not have one made of dirt. There are no cars. No subways. No four-wheel-drive buses. The name of the city is Venice.

More than 1,000 years ago, this city was *erected* on many small *islands.* The islands are in Italy in the Adriatic Sea. Waterways are the streets and avenues of the city. They are called *canals.* These canals are filled with *gondolas,* just as your streets are filled with cars. Traffic jams are everywhere. There is very little traffic noise, no horns, and no screeching stops. You can talk with someone across a canal.

The gondolas hold two to four *passengers.* The driver, or gondolier, stands on a deck at the back of the gondola. He rows, or moves, the boat with a long oar. Some people *prefer* the larger, less expensive *water-bus.* This is a motor-driven boat. It makes *scheduled* stops along the larger canals.

With all these boats moving through the busy canals, traffic must be *regulated.* There are no stoplights and stop signs at corners. Traffic police ride around in police boats. They even give out traffic *tickets.*

Can you imagine getting a ticket? You just broke the speed limit of 5 miles per hour!

UNDERSTANDING THE STORY

>>>> *Circle the letter next to each correct statement.*

1. Gondolas are more popular with tourists than the less expensive water-bus because
   a. gondoliers are friendlier than the driver of the water-bus.
   (b.) gondolas are an important part of the charm of Venice.
   c. all visitors to Venice have a lot of money to spend.

2. It is possible to get a traffic ticket in Venice if
   a. your gondola doesn't have parking lights.
   (b.) you carry more than four passengers in your gondola.
   c. you ignore the stoplights.

MAKE AN ALPHABETICAL LIST

>>>> *Here are the ten vocabulary words in the lesson. Write them in alphabetical order in the spaces below.*

erected	islands	canals	gondolas	passengers
prefer	water-bus	scheduled	regulated	tickets

1. canals
2. erected
3. gondolas
4. islands
5. passengers

6. prefer
7. regulated
8. scheduled
9. tickets
10. water-bus

WHAT DO THE WORDS MEAN?

>>>> *Following are some meanings, or definitions, for the ten vocabulary words in this lesson. Write the words next to their definitions.*

1. gondolas — long narrow boats with a high peak at each end
2. tickets — notices you get from a police officer for breaking the law
3. erected — built; constructed
4. regulated — kept in working order; controlled
5. islands — bodies of land surrounded by water
6. scheduled — happening at definite times
7. passengers — people who travel in a bus, boat, train, or plane
8. canals — waterways used like roads
9. prefer — to want one thing instead of another
10. water-bus — a canal boat with a motor that carries many passengers

22

COMPLETE THE SENTENCES

>>>> *Use the vocabulary words in this lesson to complete the following sentences. Use each word only once.*

islands	scheduled	prefer	regulated	gondolas
canals	passengers	erected	water-bus	tickets

1. The water-bus makes _____scheduled_____ stops, but the gondolas do not.

2. We took the_____water-bus_____ because we were in a hurry and low on money.

3. People hire _____gondolas_____ when they want a slow, quiet tour of the city.

4. The _____passengers_____ in different boats waved to each other.

5. Venice is made up of many small _____islands_____ connected by a series of waterways.

6. Though the water-bus costs less, most people _____prefer_____ to ride in the gondolas because they are more graceful.

7. In Venice, many statues have been _____erected_____ to honor important citizens.

8. To avoid getting _____tickets_____, the water-bus driver obeys the traffic rules.

9. The gondolas in Venice are just as carefully _____regulated_____ as taxicabs are in other cities.

10. It is a shame to see litter floating in the _____canals_____.

USE YOUR OWN WORDS

>>>> *Look at the picture. What words come into your mind other than the ten vocabulary words used in this lesson? Write them on the lines below. To help you get started, here are two good words:*

1. _____buildings_____
2. _____water_____
3. ____Answers will vary.____
4. _____
5. _____
6. _____
7. _____
8. _____
9. _____
10. _____

WHICH WORD IS NOT A SYNONYM?

>>>> A **synonym** is a word that means the same as another word.

>>>> *Below are six vocabulary words. They are followed by four other words or phrases. Three are synonyms; one is not. Circle the word or phrase that is **not** a synonym.*

Vocabulary Words		Synonyms		
1. **erected**	built	constructed	(destroyed)	created
2. **regulated**	controlled	ordered	governed	(not uniform)
3. **prefer**	(dislike)	favor	choose	like better
4. **gondolas**	boats	(autos)	watercraft	vessels
5. **passengers**	riders	travelers	(operators)	persons who ride
6. **canals**	waterways	streets of water	watercourses	(roadways)

COMPLETE THE STORY

>>>> Here are the ten vocabulary words for this lesson:

erected	islands	gondolas	passengers	prefer
water-bus	scheduled	canals	regulated	tickets

>>>> *There are five blank spaces in the story below. Five vocabulary words have already been used in the story. They are underlined. Use the other five words to fill in the blanks.*

Venice is an old city. The first buildings were _____erected_____ more than 1,000 years ago. Its canals are famous in song and story. But every spring, floods overflow the waterways. This water could someday destroy the _____islands_____ of Venice. Every hundred years, Venice sinks about a foot into the Adriatic Sea. The dirt and smoke from nearby factories are also harming its buildings. This pollution must be _____regulated_____ to save Venice.

Once, there were 10,000 gondolas. Now, there are less than 300. The people prefer the cheaper _____water-bus_____. Venice's biggest business today comes from the _____passengers_____ whose tickets let them off scheduled cruise ships and planes to see the sights.

24

Learn More About Venice

>>>> *On a separate piece of paper or in your notebook or journal, complete one or more of the activities below.*

Building Language

All languages contain idioms, or figures of speech, that are unique to them. Look up words in English that are "borrowed" from Italian. Write them down and explain what they mean.

Learning Across the Curriculum

Venice is celebrated for its art and artists. Research some of this art. Make a copy of the piece of art you like best. It can be a painting, a sculpture, or even a building. Explain something about the history of the work and its artist. Then explain why you like it.

Broadening Your Understanding

Your town has suddenly become, like Venice, a city without paved streets. Instead, all the streets are waterways. Think about how life would be different in your town. Write an imaginary journal entry about a day in your life and how your everyday routine would change.

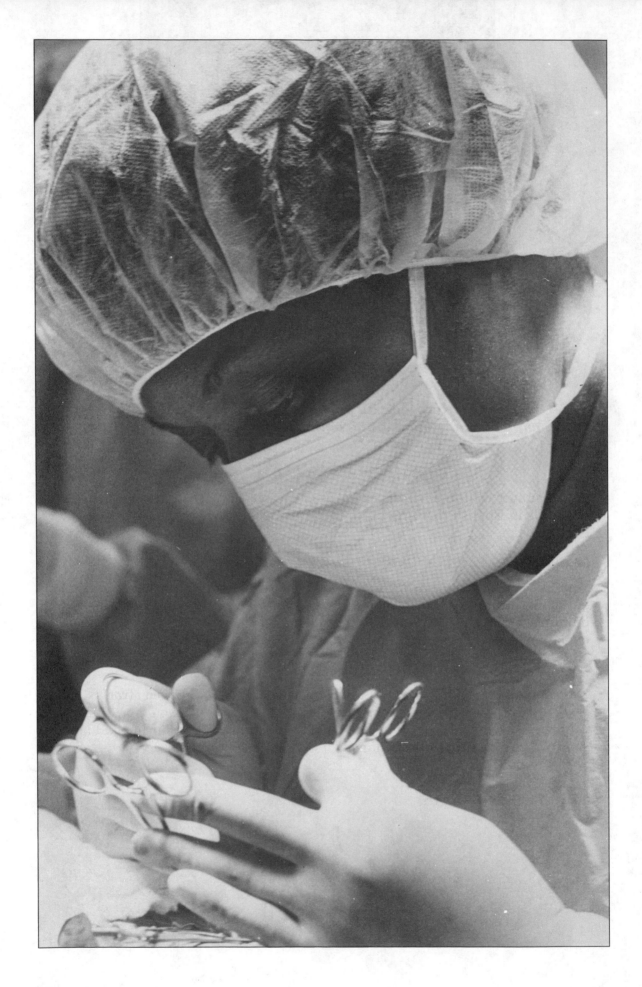

When Alexa Canady was a child, African American children had to overcome many **obstacles** before they could succeed. One of her elementary school teachers in Lansing, Michigan, where she was born in 1950, refused to believe Canady's test scores and lied about the results. Later, Canady's family moved to a house that was sold to them because they were African American. The seller wanted to "punish" his neighbors. Canady says, **"Racism** was always presented to me as *their* problem and not *our* problem."

The Civil Rights **Movement** of the early 1960s changed Canady's life. She was able to travel and to seek education at places that would not **previously** have been open to her. She says that she filled two **quotas** at once—she was African American and female.

Ultimately, Canady went to medical school and then specialized in **neurosurgery.** She became the first African American woman neurosurgeon in the United States. There were always African American doctors. They had almost **exclusively** African American patients. Canady believes that this is all beginning to change. Today people of all races go to African American doctors when they need their skills.

Canady has a sense of humor, but she is saddened by the **tasks** still to be done. She feels that people still do not have confidence in women. She wants people to overcome their **bias** against African Americans and against women. Only time will tell whether she will see a society that deals properly with racial and gender prejudice.

UNDERSTANDING THE STORY

>>>> *Circle the letter next to each correct statement.*

1. Another good title for this story might be:
   a. "How to Be an African American Doctor."
   b. "A Hard Road to Success."
   c. "Some People Don't Want Success."

2. Probably, the biggest problem faced by Canady was
   a. lack of money for school.
   b. the difficulty of traveling to school.
   c. people's prejudice about African American women doctors.

MAKE AN ALPHABETICAL LIST

>>>> *Here are the ten vocabulary words in the lesson. Write them in alphabetical order in the spaces below.*

| obstacles | ultimately | racism | neurosurgery | movement |
| exclusively | previously | tasks | quotas | bias |

1. bias
2. exclusively
3. movement
4. neurosurgery
5. obstacles
6. previously
7. quotas
8. racism
9. tasks
10. ultimately

WHAT DO THE WORDS MEAN?

>>>> *Following are some meanings, or definitions, for the ten vocabulary words in this lesson. Write the words next to their definitions.*

1. obstacles — barriers; difficulties

2. movement — crusade; organized effort

3. neurosurgery — operations on the brain, spinal cord, or nerves

4. tasks — jobs; assignments

5. previously — earlier; before

6. quotas — shares; positions held for a certain group

7. racism — a belief that one's own race is better than another's

8. exclusively — entirely; completely

9. bias — an unfair opinion or influence in favor of or against someone or something

10. ultimately — finally; at last

COMPLETE THE SENTENCES

>>>> *Use the vocabulary words in this lesson to complete the following sentences. Use each word only once.*

obstacles	ultimately	racism	neurosurgery	movement
exclusively	previously	tasks	quotas	bias

1. Alexa Canady completes many _____ tasks _____ throughout her day.
2. She thinks there are too many _____ obstacles _____ to success for African Americans and women in America.
3. She hopes that women and African Americans _____ ultimately _____ will be accepted as doctors.
4. Some people have a _____ bias _____ against African Americans and women, but she hopes that will change.
5. The Civil Rights _____ Movement _____ made her education as a doctor possible.
6. Schools had _____ quotas _____ of African Americans, women, and other minorities to fill.
7. Canady studied _____ neurosurgery _____ and practices in Detroit.
8. Discrimination does not apply _____ exclusively _____ to African Americans or women; many other minorities are involved.
   groups
9. Today laws prevent discrimination; _____ previously _____, however, minorities found it more difficult to succeed.
10. _____ Racism _____ means to judge people by skin color or racial characteristics.

USE YOUR OWN WORDS

>>>> *Look at the picture. What words come into your mind other than the ten vocabulary words used in this lesson? Write them on the lines below. To help you get started, here are two good words:*

1. _____ mask _____
2. _____ doctor _____
3. _____ Answers will vary. _____
4. _____
5. _____
6. _____
7. _____
8. _____
9. _____
10. _____

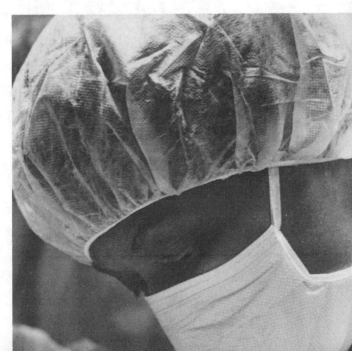

UNSCRAMBLE THE LETTERS

>>>> *Each group of letters contains the letters in one of the vocabulary words for this lesson. Can you unscramble them? Write your answers in the lines to the right of each letter group.*

Scrambled Words	Vocabulary Words
1. utosaq	quotas
2. luyletamit	ultimately
3. carmis	racism
4. evenmomt	movement
5. sbai	bias
6. kstsa	tasks
7. liseucyxvle	exclusively
8. losiprveuy	previously
9. enrouysrgeur	neurosurgery
10. slecabost	obstacles

COMPLETE THE STORY

>>>> Here are the ten vocabulary words for this lesson:

obstacles	ultimately	racism	neurosurgery	movement
exclusively	previously	tasks	quotas	bias

>>>> *There are five blank spaces in the story below. Five vocabulary words have already been used in the story. They are underlined. Use the other five words to fill in the blanks.*

It has not always been easy for minorities to succeed. After the <u>movement</u> for civil rights, many schools and businesses set _____quotas_____ so that minorities would get a fair chance at a good job. <u>Previously</u>, people thought they did not want to work with minorities. As time passed, these people learned that their _____bias_____ was not based on fact but on their own fears. They had been guilty of _____racism_____ because they were ignorant of other races.

Of course, the _____obstacles_____ for minorities are not ended. There are very few minorities in <u>neurosurgery</u>. _____Ultimately_____, people would like to see minorities in every kind of job. This work is not <u>exclusively</u> the job of minorities. We must all work to help American citizens overcome <u>obstacles</u> so they can lead better lives.

Learn More About Discrimination

>>>> *On a separate piece of paper or in your notebook or journal, complete one or more of the activities below.*

Learning Across the Curriculum

Canady says the Civil Rights Movement of the 1960s gave opportunities that helped her become a doctor. Research the Civil Rights movement. Write about the kinds of opportunties it provided for African American people in the United States.

Broadening Your Understanding

The Americans with Disabilities Act bars discrimination against persons with disabilities. The law applies to most public places, including schools. Interview the person who helps make sure this law is carried out at your school. Find out what your school has done to put the ADA into action. Then give a report to your classmates.

Extending Your Reading

Many African Americans have become scientists who made important discoveries. Read one of the following books and write a report about the scientist in whom you are most interested. Explain what he or she accomplished.

Black Pioneers of Science and Invention, by Louis Haber
Seven Black American Scientists, by Robert Hayden

6 SOUL AT ITS BEST

The lights were *dim.* The stage was empty. The people were still. Slowly, a soft *glow* from the stage lights broke through the darkness. The *hush* of the moment was broken. The announcer's *booming* voice called out "Aretha Franklin." The crowd rose to its feet. A *celebration* was about to begin. Everyone was ready for the super sister of soul. Quietly and *elegantly* Franklin came on the stage. With the *brilliance* of a sunburst, the spotlight splashed a warm glow on the star. The dark *shadows* of the empty stage were gone. Franklin's face spread light and energy over the crowd. She moved to center stage. As the crowd roared its approval, Franklin broke into a song. Sounds of her golden voice filled the room. The warm words from "Respect" and "Natural Woman" *charmed* the crowd. Then, she turned to two of her later hits, "I Can't Turn You Loose" and "Jump to It." Her version of "Amazing Grace" brought the audience to its feet once more.

The crowd's reaction was nothing new for Franklin. She has *entranced* audiences since she was a young girl. As a child, she used to sing in her father's church. Aretha performed alongside other great gospel singers. This experience had a great influence on her life. She drew upon it years later to create a unique style of singing. Her blend of gospel and blues has been popular with audiences for more than 30 years.

More recently, Franklin was featured in Whitney Houston's "How Will I Know" music video. In this video, Houston sings to Franklin the words, "I'm asking you 'cause you know about these things." This statement is a great tribute from a fellow musician.

UNDERSTANDING THE STORY

>>>> *Circle the letter next to each correct statement.*

1. The main idea of this story is that
   a. Aretha Franklin is the best singer in the business.
   b. Aretha Franklin is a very talented and well-loved singer.
   c. success does not come overnight.

2. From this story, you get the idea that
   a. Franklin has a lot to learn about playing the piano.
   b. Franklin is popular only in certain parts of the country.
   c. Franklin gives a very exciting concert.

33

MAKE AN ALPHABETICAL LIST

>>>> *Here are the ten vocabulary words in the lesson. Write them in alphabetical order in the spaces below.*

dim	shadows	glow	elegantly	charmed
brilliance	hush	celebration	booming	entranced

1. booming
2. brilliance
3. celebration
4. charmed
5. dim

6. elegantly
7. entranced
8. glow
9. hush
10. shadows

WHAT DO THE WORDS MEAN?

>>>> *Following are some meanings, or definitions, for the ten vocabulary words in this lesson. Write the words next to their definitions.*

1. brilliance — brightness; sparkle

2. booming — with a loud, deep sound

3. elegantly — gracefully; beautifully

4. glow — a light; to shine

5. charmed — pleased and delighted

6. shadows — areas of darkness or shade

7. dim — only partly lighted

8. celebration — a party in honor of something

9. entranced — filled with wonder

10. hush — quiet; sudden silence

>>>> *Use the vocabulary words in this lesson to complete the following sentences. Use each word only once.*

celebration	brilliance	entranced	booming	hush
elegantly	dim	glow	shadows	charmed

1. When the house lights became _____dim_____, it meant that the concert was about to begin.

2. Because she was in the _____shadows_____, I couldn't see the star's face.

3. With a sudden _____brilliance_____, the spotlight lighted up the whole stage.

4. Franklin _____elegantly_____ walked on the stage wearing a white sparkling gown.

5. I was so _____entranced_____ by the show, I never took my eyes off the stage.

6. A soft _____glow_____ from the footlights gave everything a cozy feeling.

7. The _____booming_____ sound of the bass guitar filled the room.

8. The concert felt like a _____celebration_____ of life and happiness.

9. Franklin's voice seemed to _____hush_____ the audience into silence.

10. As the last note faded away, the audience remained _____charmed_____ by what they had heard.

USE YOUR OWN WORDS

>>>> *Look at the picture. What words come into your mind other than the ten vocabulary words used in this lesson? Write them on the lines below. To help you get started, here are two good words:*

1. _____gown_____

2. _____microphone_____

3. ____Answers will vary.____

4. _____

5. _____

6. _____

7. _____

8. _____

9. _____

10. _____

>>>> Do you remember what a **synonym** is? It is a word that means the same or nearly the same as another word. *Sad* and *unhappy* are synonyms.

>>>> *Six of the vocabulary words for this lesson are listed below. To the right of each word are three other words. Two of them are synonyms for the vocabulary word. Draw a circle around the two synonyms for each vocabulary word.*

Vocabulary Words	*Synonyms*		
1. **celebration**	(festival)	election	(party)
2. **booming**	colorful	(loud)	(powerful)
3. **elegantly**	(gracefully)	plainly	(beautifully)
4. **hush**	(stillness)	(quiet)	strength
5. **charm**	fool	(please)	(delight)
6. **dim**	sunny	(dull)	(dark)

>>>> Here are the ten vocabulary words for this lesson:

brilliance	elegantly	hush	glow	charmed
dim	entranced	booming	celebration	shadows

>>>> *There are five blank spaces in the story below. Five vocabulary words have already been used in the story. They are underlined. Use the other five words to fill in the blanks.*

Aretha Franklin did not become a star overnight. Very few people do. When she was a girl, she sang in her father's church. Even then, people knew she was special. A _____hush_____ would come over the church members when Franklin started to sing. With the _____brilliance_____ of sunlight through an open window, her voice would ring out. She <u>charmed</u> everyone. They all said that someday the whole world might be <u>entranced</u> by her voice.

Today, when the lights _____dim_____ and the announcer's <u>booming</u> voice calls out her name, a <u>celebration</u> begins. From out of the <u>shadows</u>, an _____elegantly_____ dressed woman walks on stage. The crowd stands and applauds wildly. With the soft, warm _____glow_____ of the footlights on her face, Franklin throws kisses to the audience and begins to sing. She is certainly a star now.

Learn More About Soul Music

>>>> *On a separate piece of paper or in your notebook or journal, complete one or more of the activities below.*

Appreciating Diversity

Some African Americans say that soul music reflects their heritage. Think about music you have listened to from other cultures. Borrow several tapes or CDs from a local library and listen to them. Research how this music reflects a particular culture. Share your information and any recordings with your classmates.

Learning Across the Curriculum

Listen to some recordings of soul music. Some well-known artists are James Brown, Ray Charles, Otis Redding, and Muddy Waters. Write about how the music makes you feel. Write what you think its origins may be. Then look up a definition of soul music and compare it with your guess.

Broadening Your Understanding

Soul food is said to reflect African American heritage as much as soul music does. Research soul food and why it was developed. Recipe books are a good resource. Find a recipe you are interested in and make it for the class. Explain the history of the dish if you know it.

You don't have to go to Africa to see lions **roam** freely. You can see lions here in America. Many **species** of wild animals are kept in special parks. The animals are out in the open. They are not in cages.

This **concept** is not new. It was started by Paul Kruger in South Africa. He **developed** a park for wild animals more than 70 years ago. He was way ahead of his time. His idea has now been copied in the United States.

Visitors drive through these parks. They aren't **allowed** to get out of their cars. It would not be safe. People look out their windows at the different animals. Sometimes the animals come up to the cars.

Many animals' homes have been destroyed by careless people. But some people are concerned about animals. They are **interested** in those that are disappearing. These people **provide** money for national parks. The animals now have new homes. These parks may mean the **survival** of many **rare** wild animals.

UNDERSTANDING THE STORY

>>>> *Circle the letter next to each correct statement.*

1. Another good title for this story might be
   a. "Zoos Without Cages."
   b. "A Drive Through the Park."
   c. "Even Tame Animals Can Be Dangerous."

2. Many rare wild animals do well in an open park because
   a. people are not allowed in the park.
   b. they are fed special food and given special care.
   c. the surroundings are similar to the animals' natural surroundings.

MAKE AN ALPHABETICAL LIST

>>>> *Here are the ten vocabulary words in the lesson. Write them in alphabetical order in the spaces below.*

species	interested	allowed	visitors	provide
developed	roam	survival	concept	rare

1. allowed
2. concept
3. developed
4. interested
5. provide

6. rare
7. roam
8. species
9. survival
10. visitors

WHAT DO THE WORDS MEAN?

>>>> *Following are some meanings, or definitions, for the ten vocabulary words in this lesson. Write the words next to their definitions.*

1. allowed — let; permitted

2. roam — to move about as one pleases; to wander

3. concept — an idea; a plan

4. developed — built up; grew

5. interested — wanting to know more about something; concerned

6. provide — to give; to supply

7. rare — scarce; only a few left

8. survival — staying alive; existing

9. species — animals that have some common characteristics or qualities

10. visitors — people who visit; sightseers

COMPLETE THE SENTENCES

>>>> *Use the vocabulary words in this lesson to complete the following sentences. Use each word only once.*

roam	developed	allowed	species	visitors
interested	concept	rare	provide	survival

1. Animals that are similar in some ways belong to the same _____species_____.

2. If you are _____interested_____ in animals, you must visit the San Diego Zoo.

3. The _____survival_____ of many rare wild animals depends on our ability to provide them with new homes.

4. A _____rare_____ type of goat lives in the Andes Mountains.

5. No one is _____allowed_____ to see the newborn polar bear right away.

6. When you see animals that _____roam_____ freely, you realize how beautiful they are.

7. The _____concept_____ that a zoo does not need to have cages was started by Paul Kruger in South Africa more than 70 years ago.

8. _____Visitors_____ agree that this new type of park is a big improvement over the old-fashioned zoo.

9. The person who _____developed_____ this new type of park deserves our praise.

10. We must try to _____provide_____ the money needed to keep these parks open.

USE YOUR OWN WORDS

>>>> *Look at the picture. What words come into your mind other than the ten vocabulary words used in this lesson? Write them on the lines below. To help you get started, here are two good words:*

1. _____car_____
2. _____mane_____
3. Answers will vary.
4. _____
5. _____
6. _____
7. _____
8. _____
9. _____
10. _____

MATCH THE ANTONYMS

>>>> An **antonym** is a word that means the opposite of another word. *Fast* and *slow* are antonyms. Match the vocabulary words on the left with the antonyms on the right.

>>>> *Write the correct letter in the space next to the vocabulary word.*

	Vocabulary Words		**Antonyms**
1.	_forbidden_ **allowed**	a.	ordinary
2.	_ordinary_ **rare**	b.	forbidden
3.	_unconcerned_ **interested**	c.	unconcerned
4.	_extinction_ **survival**	d.	destroyed
5.	_destroyed_ **developed**	e.	extinction
6.	_take_ **provide**	f.	take

COMPLETE THE STORY

>>>> Here are the ten vocabulary words for this lesson:

interested	rare	provide	visitors	roam
concept	allowed	survival	species	developed

>>>> *There are five blank spaces in the story below. Five vocabulary words have already been used in the story. They are underlined. Use the other five seven words to fill in the blanks.*

What's the new and different _____concept_____ of an animal park? It's one where animals are _____allowed_____ to <u>roam</u> free! Many people find it more exciting to see wild animals on the loose instead of in cages. The parks _____provide_____ <u>visitors</u> with hours of fun. But while they have fun, the people learn about wildlife. Many Americans are _____interested_____ in the <u>survival</u> of <u>rare</u> _____species_____ of animals. We can be proud that we have <u>developed</u> this kind of park here.

Learn More About Wild Animals

>>>> *On a separate piece of paper or in your notebook or journal, complete one or more of the activities below.*

Working Together

Find out from your state natural resources department or parks department the animals that are listed as endangered species in your state. Have each person in your group choose one of the animals to research. Why did it become endangered? What is being done to save it? Have each person write a one-page report about the animal and illustrate it with a picture or photograph. Put the reports together in a book for the class.

Broadening Your Understanding

Imagine you are in charge of asking people to give money to a wild animal park, a place that protects wild animals. Write a letter or design a brochure that will convince people to give money to your wild animal park.

Extending Your Reading

Read one of the following books about animals found in the wild or choose a book about a wild animal in which you are interested. Find out about one animal's habits and needs. Design the perfect park environment for your animal.

Amazing Animals of Australia, by National Geographic
Alligators, by Patricia Lauber
Bats, by Sylvia A. Johnson
Arctic Fox, by Gail LaBonte
Apes and Monkeys, by Donald R. Shile

As a young girl, Nicholasa Mohr loved to draw. With a scrap of paper and a few crayons, she could enter an *imaginary* world. This world was filled with space and freedom.

Mohr's real world was just the opposite of her imaginary world. Her real world was marked by *poverty.* When Nicholasa was 8, her father died. She lived with her mother and six brothers in a *cramped* apartment. Her mother then became ill. She died when Mohr was in high school.

Mohr also faced discrimination because she was a Puerto Rican. She learned that Puerto Rican *females* were expected to get married and to have children. They were not expected to go to college or to work outside the home. It was *ridiculous* for her to dream of a career as an artist!

Mohr refused to give up her dreams. She attended a high school of *fashion* and *design.* After high school, she continued her education in art. She worked many different jobs to *obtain* enough money to travel. She went to Mexico to study the works of that country's great painters. The strong colors and bold designs of their works touched her deeply. Mohr felt that Mexican art somehow summed up her life as a Puerto Rican woman.

Upon returning home, Mohr set up an art studio. Her art began to be noticed for its unique design. A publisher asked Nicholasa to write a book about her *experiences.* The result was *Nilda*, a story about a Puerto Rican girl growing up in New York City. Just like the author, the main character in the book uses her imagination to escape her *dreary* surroundings.

Since *Nilda*, Mohr has written many other books about her Puerto Rican heritage. Each book has been a great success. Even as an adult, Nicholasa Mohr is still drawing—only now she uses words to create a picture of her culture!

UNDERSTANDING THE STORY

 Circle the letter next to each correct statement.

1. The statement that best expresses the main idea of this selection is:
   a. Great works of art are found in Mexico.
   b. Artists should travel to increase their skills.
   c. A female struggled to become a successful artist.

2. From this story, you can conclude that
   a. Nicholasa Mohr is determined.
   b. Nicholasa Mohr is lonely.
   c. Nicholasa Mohr likes to travel.

MAKE AN ALPHABETICAL LIST

>>>> *Here are the ten vocabulary words in the lesson. Write them in alphabetical order in the spaces below.*

imaginary	poverty	cramped	females	ridiculous
experiences	fashion	obtain	design	dreary

1. cramped
2. design
3. dreary
4. experiences
5. fashion

6. females
7. imaginary
8. obtain
9. poverty
10. ridiculous

WHAT DO THE WORDS MEAN?

>>>> *Following are some meanings, or definitions, for the ten vocabulary words in this lesson. Write the words next to their definitions.*

1. females — girls; women
2. poverty — state of being poor
3. design — forms, colors, or details arranged in a certain way
4. ridiculous — funny, silly
5. fashion — current style of dress
6. imaginary — existing only in the mind or imagination; unreal
7. experiences — all actions or events that make up a person's life
8. dreary — gloomy; dull
9. cramped — crowded; tight
10. obtain — to gain possession of

46

COMPLETE THE SENTENCES

>>>> *Use the vocabulary words in this lesson to complete the following sentences. Use each word only once.*

design	fashion	ridiculous	experiences	females
poverty	cramped	obtain	imaginary	dreary

1. Nicholasa Mohr grew up in a life of _____poverty_____.

2. Her family lived in a _____cramped_____ apartment in New York City.

3. She escaped her _____dreary_____ surroundings through her drawings.

4. Her pictures took her to an _____imaginary_____ world.

5. She studied _____fashion_____ and design in high school.

6. She discovered that people thought Puerto Rican _____females_____ should not have a career.

7. This idea was quite _____ridiculous_____ to Mohr!

8. She worked hard to _____obtain_____ the skills needed to become a great artist.

9. She studied the bold _____design_____ in the works of great Mexican artists.

10. Mohr's life _____experiences_____ are reflected in her art and writing.

USE YOUR OWN WORDS

>>>> *Look at the picture. What words come into your mind other than the ten vocabulary words used in this lesson? Write them on the lines below. To help you get started, here are two good words:*

1. _____writer_____
2. _____Puerto Rican_____
3. _____Answers will vary._____
4. _____
5. _____
6. _____
7. _____
8. _____
9. _____
10. _____

FIND THE ANTONYMS

>>>> **Antonyms** are words that are opposite in meaning. For example, *fast* and *slow* are antonyms.

>>>> *Match the vocabulary words on the left with the antonyms on the right. Write the correct letter in the space next to the vocabulary word.*

Vocabulary Words		Antonyms
1. _luxury_ **poverty**		a. spacious
2. _spacious_ **cramped**		b. real
3. _lose_ **obtain**		c. males
4. _males_ **females**		d. bright
5. _bright_ **dreary**		e. lose
6. _real_ **imaginary**		f. luxury

COMPLETE THE STORY

>>>> Here are the ten vocabulary words for this lesson:

cramped	fashion	ridiculous	experiences	imaginary
females	poverty	dreary	obtain	design

>>>> *There are five blank spaces in the story below. Five vocabulary words have already been used in the story. They are underlined. Use the other five words to fill in the blanks.*

Today, Nicholasa Mohr is a successful writer. Many of her books tell the story of Puerto Rican <u>females</u>. The stories are generally set in _____dreary_____ surroundings. The characters live a life of <u>poverty</u>. Mohr's books are based upon her own life _____experiences_____. She, too, grew up in a <u>cramped</u> home. Her family did not have enough money to _____obtain_____ fancy things.

Mohr worked hard to become a success. She did not think her dream of becoming famous was <u>ridiculous</u>. She studied _____fashion_____ and <u>design</u> in high school. She attended special art classes. She went to Mexico to study the works of great artists. The hard work paid off! Today, Nicholasa Mohr is a famous author. Through much effort, her _____imaginary_____ world has become her real world.

Learn More About Nicholasa Mohr

>>>> *On a separate piece of paper or in your notebook or journal, complete one or more of the activities below.*

Broadening Your Understanding

Nicholasa Mohr used crayons to create an imaginary world. Make a drawing of an imaginary world that you would like to visit. On the back of your drawing, explain how your imaginary world is different from the real world.

Learning Across the Curriculum

Mohr's family moved from Puerto Rico to the United States. Use reference books to learn more about Mohr's native land. Look for information about other famous Americans who came from Puerto Rico, such as Raul Julia or Tito Puente. Share your findings in a report to your class.

Appreciating Diversity

Nicholasa Mohr traveled to Mexico to study the works of great Mexican artists. Learn more about the unique style of these painters. Find out how they used their art to express their feelings about their country. Write an art review to share your ideas about what you have learned. Display pictures of some of the artists' works to help illustrate your review.

9 A DANGEROUS JUMP

Think about the *various* ways that people get to their jobs. Some drive cars to work. Others ride trains or even walk to their jobs. Some people, however, parachute to work! These people are smokejumpers, a unique group of firefighters.

Smokejumpers make up a *division* of the U.S. Forest Service. The service set up the group about 50 years ago. Its purpose was to battle forest fires in the Rocky Mountains. The slopes of the mountains are very steep. Reaching the fires by foot is very difficult. The only way to get the firefighters to the fire is to drop them at the scene by *helicopter.* These firefighters, now called smokejumpers, parachute to their job!

Work as a smokejumper has its *risks.* Weather conditions can change suddenly. A small fire can become a wall of flames. This is exactly what happened in July 1994. A group of smokejumpers were battling a fire in the mountains of Colorado. The smokejumpers seemed to have the fire under control. Suddenly, *gusts* of wind roared through the area. The fire exploded into a firestorm. The storm of heat *advanced* at a *rate* of 100 feet per minute!

The smokejumpers raced for their lives. Some were unable to outrun the wall of fire. Others *sought* protection inside foil blankets called fire shelters. Still, 14 smokejumpers died in the fire. They died trying to find escape routes through parts of the forest that had burned.

By the next day, the fire had been *smothered.* The smokejumpers had done their job. They reviewed what had happened to make sure it would not happen again. Then they got ready for their next battle with *nature.*

UNDERSTANDING THE STORY

>>>> *Circle the letter next to each correct statement.*

1. The statement that best expresses the main idea of this selection is:
   a. Smokejumping is a dangerous job.
   b. People get to work in many different ways.
   c. Parachute diving is fun.

2. From this story, you can conclude that
   a. fires rarely happen in the Rocky Mountains.
   b. it is easier to fight a fire on a calm day than on a windy day.
   c. it is impossible to stop forest fires that burn on mountainsides.

MAKE AN ALPHABETICAL LIST

>>>> **Here are the ten vocabulary words in the lesson. Write them in alphabetical order in the spaces below.**

various	nature	division	smothered	gusts
risks	sought	rate	helicopter	advanced

1. advanced
2. division
3. gusts
4. helicopter
5. nature

6. rate
7. risks
8. smothered
9. sought
10. various

WHAT DO THE WORDS MEAN?

>>>> **Following are some meanings, or definitions, for the ten vocabulary words in this lesson. Write the words next to their definitions.**

1. risks — hazards; possibilities of danger

2. nature — the natural world

3. division — section; group

4. sought — searched for; tried to find

5. various — different kinds; more than one

6. smothered — to cut off oxygen supply; suffocated

7. advanced — moved forward

8. rate — measured quantity

9. gusts — violent rushes; sudden outbursts

10. helicopter — aircraft with circular blades

52

COMPLETE THE SENTENCES

>>>> *Use the vocabulary words in this lesson to complete the following sentences. Use each word only once.*

gusts	sought	helicopter	smothered	risks
rate	various	division	nature	advanced

1. Smokejumpers make up a _____division_____ of the U.S. Forest Service.

2. They reach forest fires by jumping from a _____helicopter_____.

3. They face many _____risks_____ when fighting forest fires.

4. In 1994, the smokejumpers _____smothered_____ a fire in Colorado.

5. Sudden _____gusts_____ of wind made the fire turn into a wall of flames.

6. A wall of fire _____advanced_____ quickly toward the firefighters.

7. The fire moved at a _____rate_____ of 100 feet per minute.

8. The Smokejumpers _____sought_____ the safety of their fire shelters.

9. Smokejumpers use _____various_____ techniques to put out fires.

10. They saw how _____nature_____ can sometimes be very unpredictable.

USE YOUR OWN WORDS

>>>> *Look at the picture. What words come into your mind other than the ten vocabulary words used in this lesson? Write them on the lines below. To help you get started, here are two good words:*

1. _____courageous_____
2. _____dedicated_____
3. _____Answers will vary._____
4. _____
5. _____
6. _____
7. _____
8. _____
9. _____
10. _____

FIND THE ANALOGIES

>>>> In an **analogy**, similar relationships occur between words that are different. For example, *hammer* is to *carpentry* as *piano* is to *music*. The relationship is an object to its use. Here's another analogy: *rung* is to *ladder* as *seat* is to *chair*. In this relationship, the words show a part-to-whole relationship.

>>>> *See if you can complete the following analogies. Circle the correct word or words.*

1. **Computer** is to **communication** as **helicopter** is to
   a. airport   (b.) transportation   c. agriculture   d. music

2. **Splashes** are to **water** as **gusts** are to
   a. hills   b. leaves   c. waves   (d.) wind

3. **Buildings** are to **urban** as **trees** are to
   (a.) nature   b. streets   c. space   d. rivers

4. **Committee** is to **club** as **division** is to
   a. crowd   b. football   c. meetings   (d.) army

5. **Room deodorizer** is to **freshened** as **fire extinguisher** is to
   a. located   b. hope   (c.) smothered   d. flamed

COMPLETE THE STORY

>>>> Here are the ten vocabulary words for this lesson:

nature	sought	helicopter	advanced	risks
gusts	division	smothered	various	rate

>>>> *There are five blank spaces in the story below. Five vocabulary words have already been used in the story. They are underlined. Use the other five words to fill in the blanks.*

The Colorado fire of 1994 was not the first time smokejumpers lost their lives while battling <u>nature</u>. This _____division_____ of the U.S. Forest Service also lost some members in 1949. They were fighting a forest fire in Mann Gulch, Montana. Just like the Colorado incident, this fire was along the steep slopes of the Rocky Mountains. In order to reach the blaze, the smokejumpers had to be dropped by _____heliocopter_____ . As the firefighters <u>sought</u> to contain the fire, the wind suddenly picked up. The _____gusts_____ of wind fanned the fire. Suddenly, a wall of flames <u>advanced</u> at an alarming <u>rate</u> toward the smokejumpers. Before the flames were finally <u>smothered</u>, 13 smokejumpers were killed. _____Various_____ agencies reviewed what happened at Mann Gulch that day. Changes were made in the way smokejumpers attack a fire. Yet, there are certain _____risks_____ that will always be present when fighting a fire.

54

Learn More About Fighting Fires

>>>> *On a separate piece of paper or in your notebook or journal, complete one or more of the activities below.*

Learning Across the Curriculum

Smokejumpers are called in to battle fires that blaze along steep mountainsides. Look at a topographical map of the United States. Identify five states that would probably need the help of smokejumpers to put out forest fires.

Broadening Your Understanding

Your school has regular fire drills to practice leaving the building in case of a fire. Think about the design of your home. What escape routes should your family use if a fire ever breaks out in your home? Talk about this with other members of your family. Then have a family fire drill to be sure everyone knows the best way of leaving your home.

Learning Across the Curriculum

The forest fires that occurred in Colorado and Montana were started when lightning struck a dry area of ground. However, many forest fires begin because of the mistakes humans make. Make a list of things people can do to prevent forest fires. Select one item on your list to use for a poster on fire safety.

"I don't like the sound of my voice. Nor was I ever a professional athlete." Not many TV sports `announcers` would dare say that. But Bryant Gumbel did. He's very `honest.` Maybe that is why he is considered one of the best in his `field.`

Gumbel's first job in television was at a station in Los Angeles. He was the sports announcer for the weekend games. The station asked him to change his name. They thought that Gumbel sounded too much like `mumble` and `bumble.` But he refused. He took `pride` in being honest, even about his name.

Three years later, NBC in New York wanted him. He became a `key` person on the "Grandstand" show. Before long, he was the `host` for other National Football League shows. He did the shows before, between, and after the games. This isn't an easy job. No matter how many different things may be happening off camera, the on-camera announcer must look cool and calm. He or she must also make the reports interesting. Gumbel says, "The trick is not to give one audience the same information twice." This comment means that the announcer must have many different things to say.

People like Gumbel's `style.` He was very popular as a guest on the "Today" show. It wasn't long before NBC asked him to be one of the hosts of the show. Many people consider this to be one of the top jobs in television. When people say that to Gumbel, he just smiles and says he was lucky. There's that `appealing` honesty again.

UNDERSTANDING THE STORY

>>>> *Circle the letter next to each correct statement.*

1. The main idea of this story is that
   a. sports announcing is a hard job.
   b. Bryant Gumbel is a lucky person.
   c. Gumbel's honest style has brought him success.

2. Though it doesn't say so, from the story you can tell that
   a. many people would like to host the "Today" show.
   b. Gumbel had many other jobs before he broke into television.
   c. Gumbel plays football in his spare time.

MAKE AN ALPHABETICAL LIST

>>>> *Here are the ten vocabulary words in the lesson. Write them in alphabetical order in the spaces below.*

| honest | announcers | pride | key | host |
| field | mumble | style | appealing | bumble |

1. announcers
2. appealing
3. bumble
4. field
5. honest

6. host
7. key
8. mumble
9. pride
10. style

WHAT DO THE WORDS MEAN?

>>>> *Following are some meanings, or definitions, for the ten vocabulary words in this lesson. Write the words next to their definitions.*

1. mumble — to speak in an unclear way with the lips not open enough

2. announcers — people who introduce or tell about the action on TV or radio shows

3. field — the type of job one does

4. host — the main announcer on a show; someone giving a party for invited guests

5. bumble — to act in a clumsy way

6. style — a way of doing things

7. key — most important; central

8. honest — truthful; not phony

9. pride — good feelings about yourself; self-respect

10. appealing — likable; pleasing

COMPLETE THE SENTENCES

>>>> *Use the vocabulary words in this lesson to complete the following sentences. Use each word only once.*

announcers	field	key	appealing	pride
bumble	honest	host	mumble	style

1. Bryant Gumbel takes _____ pride _____ in who he is and what he does.

2. Like many other TV _____ announcers _____, he reported on sports events.

3. People said that they liked his _____ style _____, or way of doing things.

4. There is something very _____ appealing _____ about his down-to-earth, friendly ways.

5. Before long, he was asked to be the _____ host _____ of a popular TV show.

6. Very quickly, he had become one of the best in his _____ field _____.

7. He had to be careful not to _____ mumble _____ if he wanted people to understand what he was saying.

8. If he were to _____ bumble _____ through every report, he would be laughed right out of the business.

9. Because Gumbel was the _____ key _____ person on "Grandstand," he had to be present for every show.

10. Gumbel would agree that it pays to be _____ honest _____ rather than phony.

USE YOUR OWN WORDS

>>>> *Look at the picture. What words come into your mind other than the ten vocabulary words used in this lesson? Write them on the lines below. To help you get started, here are two good words:*

1. _____ speaking _____
2. _____ tie _____
3. _____ Answers will vary. _____
4. _____
5. _____
6. _____
7. _____
8. _____
9. _____
10. _____

CIRCLE THE SYNONYMS

>>>> Do you remember what a **synonym** is? It is a word that means the same or nearly the same as another word. *Sad* and *unhappy* are synonyms.

>>>> *Six of the vocabulary words for this lesson are listed below. To the right of each word are three words or phrases. Two of them are synonyms for the vocabulary word. Draw a circle around the two synonyms for each vocabulary word.*

Vocabulary Words   **Synonyms**

1. **honest**   (truthful)   handy   (sincere)
2. **key**   loud   (most   (essential)
                                      important)
3. **field**   (line of work)   (specialty)   neighborhood
4. **appealing**   (attractive)   lazy   (likable)
5. **style**   hope   (manner)   (approach)
6 **host**   (announcer)   guest   (master of
                                                     ceremonies)

COMPLETE THE STORY

>>>> Here are the ten vocabulary words for this lesson:

announcers	field	honest	pride	bumble
style	host	key	mumble	appealing

>>>> *There are five blank spaces in the story below. Five vocabulary words have already been used in the story. They are underlined. Use the other five words to fill in the blanks.*

You may have noticed that people who take _____pride_____ in themselves are often very <u>appealing</u> to others. Bryant Gumbel is one of these people. He didn't let the fact that his last name rhymes with _____mumble_____ and <u>bumble</u> bother him. He knew that sports _____announcers_____ tell people who are not actually at a sports event what they need to know. For this reason, Gumbel knew that being <u>honest</u> was the best thing he could do. He was right. Many people like him because they feel that they can trust him. They like his _____style_____, or way of doing things. He is also interesting to listen to and seems like someone you would like to know. These qualities make him a good television news _____host_____. It is more than luck that has made Gumbel a <u>key</u> person in his <u>field</u>.

60

Learn More About Broadcasting

>>>> *On a separate piece of paper or in your notebook or journal,
complete one or more of the activities below.*

Building Language

Watch a sports newscast. Write down the phrases and words you do not
understand. Find out what each means and write it in a sentence. Have a friend
check your work to see if you understood what each phrase or word means.

Working Together

Watch a television talk show. Then stage your own talk show with a group of
people. You will need a director, a host, guests, and writers for the questions and
introductions. If possible, you may want to have one student film the talk show
with a video camera so the group can watch it later.

Broadening Your Understanding

Watch "60 Minutes" or another television magazine show. What kind of research
must an interviewer do in order to have a successful show? Pick a person you
would like to interview. Do research to come up with questions that will get the
information you want. Tell your class why you are interested in interviewing this
person and share with them your list of questions.

11 PRIMA BALLERINA

Maria Tallchief is an $\boxed{accomplished}$ ballerina. She has brought $\boxed{pleasure}$ to audiences around the world. $\boxed{Observers}$ have seen her leap and glide across a stage. She can $\boxed{express}$ her feelings in dance.

Tallchief was born in 1925 in Fairfax, Oklahoma. Her father was a Native American. When she was a $\boxed{youngster,}$ the family moved to California. There she began her $\boxed{preparation}$ for stardom. She studied music. She trained in ballet. Some of her teachers were famous $\boxed{professional}$ dancers.

Tallchief was a good learner. She became a $\boxed{brilliant}$ dancer. She was the star dancer of the New York City Ballet. She was also a $\boxed{guest}$ with other dance companies. She became known for her great $\boxed{attention}$ to detail.

In 1946, Tallchief married George Balanchine. He created many ballets. The two artists worked together. She danced the leading role in his ballet *The Firebird*. This role made her famous around the world.

In 1980, Tallchief started the Chicago City Ballet. She taught young dancers. She supervised the shows. Newcomers learned why she is the most brilliant American ballerina of her time.

UNDERSTANDING THE STORY

 Circle the letter next to each correct statement.

1. Maria Tallchief is known as a brilliant ballerina because she
   a. married a choreographer.
   b. is a Native American.
   c. pays great attention to detail.

2. Maria Tallchief probably
   a. wishes she could leave the world of ballet.
   b. enjoys dance even though it is hard work.
   c. would tell young ballerinas to study something else.

MAKE AN ALPHABETICAL LIST

>>>> *Here are the ten vocabulary words in the lesson. Write them in alphabetical order in the spaces below.*

professional	accomplished	guest	express	brilliant
observes	youngster	preparation	pleasure	attention

1. accomplished
2. attention
3. brilliant
4. express
5. guest

6. observers
7. pleasure
8. preparation
9. professional
10. youngster

WHAT DO THE WORDS MEAN?

>>>> *Following are some meanings, or definitions, for the ten vocabulary words in this lesson. Write the words next to their definitions.*

1. guest — visitor

2. youngster — child

3. attention — concentration

4. preparation — readiness

5. professional — having to do with earning a living in a job that requires certain skills

6. observers — viewers

7. pleasure — delight; enjoyment

8. express — to make known

9. brilliant — splendid; magnificent

10. accomplished — skilled; experienced

64

COMPLETE THE SENTENCES

>>>> *Use the vocabulary words in this lesson to complete the following sentences. Use each word only once.*

pleasure	attention	accomplished	preparation	professional
observers	guest	express	youngster	brilliant

1. Maria Tallchief is an _____accomplished_____ ballerina.

2. Her _____brilliant_____ dance delights audiences.

3. Tallchief studied music and dance as a _____youngster_____.

4. Some of her teachers were _____professional_____ dancers.

5. They taught her how to _____express_____ emotions through dance.

6. This _____preparation_____ helped Tallchief become a star.

7. She is known for paying close _____attention_____ to detail.

8. Tallchief performed as a _____guest_____ with many dance companies.

9. _____Observers_____ wonder at the way Tallchief moved her body.

10. Her dance brought great _____pleasure_____ to all members of the audience.

USE YOUR OWN WORDS

>>>> *Look at the picture. What words come into your mind other than the ten vocabulary words used in this lesson? Write them on the lines below. To help you get started, here are two good words:*

1. _____Native American_____
2. _____ballerina_____
3. _____Answers will vary._____
4. _____
5. _____
6. _____
7. _____
8. _____
9. _____
10. _____

IDENTIFY THE ANTONYMS AND SYNONYMS

>>>> There are six vocabulary words listed below. To the right of each is either a synonym or an antonym. Remember: a **synonym** is a word that means the same or nearly the same as another word. An **antonym** is a word that means the opposite of another word.

>>>> *On the line beside each pair of words, write **S** for synonyms or **A** for antonyms.*

	Vocabulary Words		*Antonyms and Synonyms*	
1.	**observers**	viewers	1.	S
2.	**pleasure**	disgust	2.	A
3.	**youngster**	adult	3.	A
4.	**brilliant**	dull	4.	A
5.	**guest**	visitor	5.	S
6.	**professional**	amateur	6.	A

COMPLETE THE STORY

>>>> Here are the ten vocabulary words for this lesson:

attention	accomplished	brilliant	professional	observers
youngster	preparation	pleasure	guest	express

>>>> *There are five blank spaces in the story below. Five vocabulary words have already been used in the story. They are underlined. Use the other five words to fill in the blanks.*

Maria Tallchief was a <u>professional</u> dancer. _____Observers_____ delighted at watching her move across a stage. The movement of her body could _____express_____ different emotions. Her dancing brought great _____pleasure_____ to every audience.

Tallchief worked hard on her dance. She spent long hours in <u>preparation</u> for a new role. She was known for her _____attention_____ to detail. Even as a <u>youngster</u>, she was devoted to ballet. Because of this effort, Tallchief became a <u>brilliant</u> dancer. She was invited to be a _____guest_____ in theaters around the world. Today, Maria Tallchief is known as an <u>accomplished</u> American ballerina.

Learn More About Ballet

>>>> *On a separate piece of paper or in your notebook or journal, complete one or more of the activities below.*

Learning Across the Curriculum

Some ballets tell a story through dance. Write or tell a story that you think could be told without words. Write words that describe the movements of the dancers in your ballet.

Broadening Your Understanding

Two other great American ballerinas of the mid-1900s were Melissa Hayden and Nora Kaye. Use reference materials to learn about the lives of these two dancers. Then, based upon what you discover, write a Help Wanted ad for a prima ballerina. Make sure to include a description of the job as well as job requirements.

Learning Across the Curriculum

It is believed that ballet began during the 1500s. Use reference materials to learn more about the history of this form of dance. Discover where and when the first ballet was performed. Create a poster that advertises this performance.

12 DISCOVERING A NEW WORLD

The captain of the fishing boat stops the engines. He looks out over the calm, clear water. A person is resting in a chair out on the deck. The mate walks over. "It's time to get ready," he says. Suddenly, you realize the person on board is . . . you!

You are about to enter a brand new world—the sea. Fish can **breathe** underwater, but people cannot. Divers must carry their own air. The captain brings over an **aqualung.** The aqualung has one or two tanks. The tanks are filled with oxygen. A rubber hose carries the oxygen to the **mouthpiece.** It is through this mouthpiece that you will breathe underwater.

You strap large **fins** to your feet. Next comes the face mask. It is made of rubber with a glass face plate. Your diving partner is dressed the same way. The captain asks if both of you are ready.

You enter the sea. Below, brightly colored **coral** looks like stone plants. Seaweed and other plants wave in the ocean current. **Schools** of tiny fish swim all around you. What's that? A shark? No, it's only a big, old grouper. These are all the true **inhabitants** of the sea. You are only a **scuba** diver.

Your experienced partner watches the time. It is passing swiftly. You don't want to run out of air. Watch your **depth** at all times. Going too deep can cause injury or death. **Surfacing** too fast can also be dangerous. All too soon, it's time to go back.

Let's do it again. But please remember: You must never dive alone.

UNDERSTANDING THE STORY

>>>> *Circle the letter next to each correct statement.*

1. Another good title for this story might be:
   a. "The True Inhabitants of the Sea."
   b. "Diving Beneath the Sea."
   c. "How to Prevent Drowning."

2. The warning "never dive alone" is probably made because
   a. it is against the law to dive alone.
   b. it is more fun to dive with a friend.
   c. a diver could run into trouble and need help.

MAKE AN ALPHABETICAL LIST

>>>> *Here are the ten vocabulary words in the lesson. Write them in alphabetical order in the spaces below.*

inhabitants	schools	fins	coral	depth
breathe	aqualung	mouthpiece	scuba	surfacing

1. aqualung
2. breathe
3. coral
4. depth
5. fins

6. inhabitants
7. mouthpiece
8. schools
9. scuba
10. surfacing

WHAT DO THE WORDS MEAN?

>>>> *Following are some meanings, or definitions, for the ten vocabulary words in this lesson. Write the words next to their definitions.*

1. inhabitants — persons or animals who live in a place

2. scuba — gear that allows breathing underwater; self-contained underwater breathing apparatus

3. fins — rubber flippers that people wear on their feet to swim and dive

4. aqualung — a dive tank that supplies air

5. mouthpiece — a part of the scuba equipment that the diver holds in the mouth

6. breathe — to take air in and let air out

7. schools — large numbers of fish swimming together

8. coral — a hard substance made by the skeletons of tiny sea animals

9. depth — the distance from top to bottom

10. surfacing — coming to the top of the water

COMPLETE THE SENTENCES

>>>> *Use the vocabulary words in this lesson to complete the following sentences. Use each word only once.*

breathe	depth	schools	inhabitants	mouthpiece
scuba	surfacing	aqualung	fins	coral

1. _____Scuba_____ stands for "<u>s</u>elf-<u>c</u>ontained <u>u</u>nderwater <u>b</u>reathing <u>a</u>pparatus."

2. Since the invention of the _____aqualung_____, people have been able to explore the world of the sea.

3. The true _____inhabitants_____ of the sea can breathe underwater.

4. Although _____coral_____ can be beautiful, its sharp edges can be dangerous.

5. When the diver found it difficult to _____breathe_____, she signaled for help.

6. _____Schools_____ of tiny green fish swam near the old shipwreck.

7. If your _____mouthpiece_____ is not firmly in place, you may swallow some water.

8. When _____surfacing_____, you must be careful not to come up too fast.

9. The _____fins_____ that you wear on your feet help you to swim.

10. When the diver reached a _____depth_____ of 70 feet, he decided he had gone deep enough.

USE YOUR OWN WORDS

>>>> *Look at the picture. What words come into your mind other than the ten vocabulary words used in this lesson? Write them on the lines below. To help you get started, here are two good words:*

1. _____bubbles_____
2. _____face mask_____
3. ____Answers will vary.____
4. _____
5. _____
6. _____
7. _____
8. _____
9. _____
10. _____

UNSCRAMBLE THE LETTERS

>>>> Each group of letters contains the letters in one of the vocabulary words for this lesson. Can you unscramble them? Write your answers in the lines to the right of each letter group.

Scrambled Words	Vocabulary Words
1. bcsua	scuba
2. lcrao	coral
3. tdhpe	depth
4. lsshcoo	schools
5. hebtera	breathe
6. carfinsug	surfacing
7. nifs	fins
8. thanbitanis	inhabitants
9. pomiethuce	mouthpiece
10. laguquna	aqualung

COMPLETE THE STORY

>>>> Here are the ten vocabulary words for this lesson:

schools	inhabitants	coral	depth	fins
breathe	mouthpiece	aqualung	scuba	surfacing

>>>> There are five blanks in the story below. Five vocabulary words have already been used in the story. They are underlined. Use the other five words to fill in the blanks.

The world beneath the sea is beautiful but dangerous. You may learn how to be one of the part-time ___inhabitants___ of the sea. But special scuba equipment and training are needed. First, you have to have a mask, fins, and an ___aqualung___. The mouthpiece of the aqualung must fit comfortably in your mouth so that you will be able to ___breathe___ underwater. There will be many lessons on how to use this equipment. You will learn to what ___depth___ you can safely descend. The instructor will teach you surfacing skills. After your training, you, too, will see ___schools___ of fish and bright coral on the bottom of the sea. Happy diving!

>>>> *On a separate piece of paper or in your notebook or journal, complete one or more of the activities below.*

Learning Across the Curriculum

The Great Barrier Reef in Australia is one of the best places in the world to scuba dive. Research the Great Barrier Reef. Find out why divers travel from around the world to dive there. What is someone who dives there likely to see?

Broadening Your Understanding

Many shops that sell diving equipment are staffed by people who love to scuba dive. Call one of these shops and interview an experienced diver. Ask about his or her most memorable dive. Write a newspaper article about your findings.

Extending Your Reading

Read one of the following books about scuba diving. Pretend you are a scuba diver on a famous dive. Write about what you saw and experienced.

Swimming and Scuba Diving, by Michael Jay
Divers, by Kendall McDonald
Sport Diving, by Carole S. Briggs

Billy Crystal was raised near New York City. He worked as a teacher. Then he decided to try to go into show business. He took care of his baby daughter during the day. At night, he went into New York City to try out his act at **comedy** clubs. He wasn't paid, but he learned how to make the people laugh. **Eventually,** he moved to California. There, he worked in television. He soon became better known. His **skills** grew, too.

Then came Crystal's big second chance. He was **hired** for the new season of "Saturday Night Live." On that show, he became **famous** for his unusual humor. The show was a perfect place for him to show off the funny **characters** that he can play. In a way, it's **ironic** that Crystal found success on the show. He was supposed to be on the original "Saturday Night Live" in 1975. At the last minute, his part was **canceled.** Crystal had to wait a few more years to get his shot at stardom.

Since leaving the show, Crystal has appeared in a number of hit movies. *City Slickers I* and *II* and *When Harry Met Sally* were very popular with his **fans.** In 1990, Crystal hosted the Academy Award ceremonies. The show was seen by nearly one billion people around the world! The viewers discovered that Billy Crystal is a very talented **comedian.**

 Circle the letter next to each correct statement.

1. Billy Crystal was
   a. on the original "Saturday Night Live."
   b. taken off the "Saturday Night Live" show at the last moment.
   c. never on "Saturday Night Live."

2. To become a good comic,
   a. you need talent but not much practice.
   b. you shouldn't practice too much.
   c. you need both talent and plenty of practice.

MAKE AN ALPHABETICAL LIST

>>>> *Here are the ten vocabulary words in the lesson. Write them in alphabetical order in the spaces below.*

characters	eventually	ironic	skills	canceled
hired	comedy	famous	fans	comedian

1. canceled
2. characters
3. comedian
4. comedy
5. eventually

6. famous
7. fans
8. hired
9. ironic
10. skills

WHAT DO THE WORDS MEAN?

>>>> *Following are some meanings, or definitions, for the ten vocabulary words in this lesson. Write the words next to their definitions.*

1. characters — people in stories, plays, and television shows and movies.

2. ironic — something that is opposite to what you would expect

3. canceled — stopped; done away with

4. comedy — something funny

5. fans — people who are enthusiastic about a performer

6. eventually — finally; in the end

7. skills — abilities that come from practice

8. hired — given a job

9. famous — well-known by many people

10. comedian — someone who tells jokes or performs in funny ways

76

COMPLETE THE SENTENCES

>>>> *Use the vocabulary words in this lesson to complete the following sentences. Use each word only once.*

characters	eventually	ironic	skills	canceled
hired	comedy	famous	fans	comedian

1. It took a long time, but _____eventually_____ Billy Crystal became a star.

2. The show was _____canceled_____ because not enough people watched it.

3. It's _____ironic_____ that Crystal did so well in a show that had once dropped him.

4. Billy Crystal does many funny _____characters_____.

5. Only if you practice will you develop your _____skills_____ as a singer.

6. I was _____hired_____ by Mr. Gomez to work in the restaurant.

7. Some people would rather see a _____comedy_____ than a serious play.

8. We were excited to see the _____famous_____ actor.

9. The _____fans_____ began to applaud the show.

10. Billy Crystal is a well-known _____comedian_____.

USE YOUR OWN WORDS

>>>> *Look at the picture. What words come into your mind other than the ten vocabulary words used in this lesson? Write them on the lines below. To help you get started, here are two good words:*

1. _____appealing_____
2. _____humor_____
3. _____Answers will vary._____
4. _____
5. _____
6. _____
7. _____
8. _____
9. _____
10. _____

UNSCRAMBLE THE LETTERS

>>>> *Each group of letters represents one of the vocabulary words for this lesson. Can you unscramble them? Write your answers in the blanks on the right.*

Scrambled Words	Vocabulary Words
1. hatescracr	characters
2. niiorc	ironic
3. yvelaluten	eventually
4. moydce	comedy
5. lcdeacne	canceled
6. nfas	fans
7. uofsam	famous
8. dimconae	comedian
9. eidhr	hired
10. lkslis	skills

COMPLETE THE STORY

>>>> Here are the ten vocabulary words for this lesson:

characters	eventually	ironic	skills	canceled
hired	comedy	famous	fans	comedian

>>>> *There are five blank spaces in the story below. Five vocabulary words have already been used in the story. They are underlined. Use the other five words to fill in the blanks.*

The _____characters_____ that he plays are sometimes different from the real Billy Crystal. The fans laugh when Crystal plays a funny single man. They can't tell that he is really a family man. Crystal is close to his children. A comedian is away from home a lot. Even so, Crystal's marriage is happy.

There are other things people may not see about Billy Crystal. He keeps trying. His big break on "Saturday Night Live" was canceled. That didn't stop him. Crystal is _____famous_____ today. But he worked hard to improve his _____skills_____. It took years before he _____eventually_____ became successful. It's ironic that comedy is such hard work. Yet, it has to look easy. You don't get _____hired_____ for good jobs until being funny no longer looks like work.

Learn More About Comedians

>>>> *On a separate piece of paper or in your notebook or journal, complete one or more of the activities below.*

Building Language

Watch a stand-up comedy routine or a situation comedy on TV. List the phrases or jokes that you do not understand. Write down your guess of what the phrase or joke means as it was used. Ask a friend to check if you are right.

Learning Across the Curriculum

Comic strips often tell jokes in an effort to look at the world in a different way. Find a comic strip from a newspaper or magazine that makes a joke. Write how the comic-strip writer uses humor. Share the information with your class.

Broadening Your Understanding

Some comedy on TV is funny. Some isn't. Watch an episode of the top-rated and bottom-rated situation comedies on TV. What makes one successful and the other not? Which did you think was funnier and why? Write reviews of the two shows.

Siberia is *bleak* and *barren.* The climate is *severe.* The winters are long. There are many blizzards. The wind blows constantly. Sometimes it reaches 100 miles per hour. The temperatures are bitter cold. The summer is short, and it's very hot.

Siberia lies in the northern part of Asia. For years, Siberia was part of the Soviet Union. Then, in 1992, this vast country split into smaller countries. Siberia became part of the Russian Federation.

More than 32 million people live in Siberia. However, there are few cities in this huge land. The towns are far apart. You can travel for days and *rarely* see anyone! Siberia's harsh climate is not good for farming. So there are few farms. However, the earth beneath the surface is rich. Iron ore, gold, silver, and other *precious* metals are mined. Oil and natural gas wells sink deep into the ground. These resources are important to the Russian *economy.*

Siberia has an unusual history. Beginning in the 18th *century,* Siberia became used as a prison *colony.* First, only criminals were sent there. Later, it became a place of *exile.* Citizens who spoke against the government were sent to work in the mines. Life in a Siberian *penal* colony usually meant an early death.

About 100 years ago, a railroad was built across Russia. This railroad helped new settlers reach Siberia. The government encouraged people to move there. They were needed to work in the mines and in the oil fields. To date, few people have moved there. Siberia may never be a land of towns and cities.

UNDERSTANDING THE STORY

 Circle the letter next to each correct statement.

1. The main idea of this story is that
   a. Siberia is best known as a prison colony.
   b. Siberia is the coldest place on earth.
   c. Siberia is bleak and barren but important to the Soviet economy.

2. Most likely, not many people have chosen to move to Siberia because
   a. the government has not offered them enough money.
   b. the riches there cannot make up for the hard life.
   c. they don't want to live among the prisoners.

MAKE AN ALPHABETICAL LIST

>>>> *Here are the ten vocabulary words in the lesson. Write them in alphabetical order in the spaces below.*

barren	rarely	precious	economy	colony
century	exile	penal	severe	bleak

1. barren
2. bleak
3. century
4. colony
5. economy

6. exile
7. penal
8. precious
9. rarely
10. severe

WHAT DO THE WORDS MEAN?

>>>> *Following are some meanings, or definitions, for the ten vocabulary words in this lesson. Write the words next to their definitions.*

1. bleak — dreary; swept by winds

2. economy — money, goods, and services

3. rarely — not often; seldom

4. penal — involving punishment

5. colony — a settlement or town set up by a group of people

6. exile — a forced removal from one's homeland

7. precious — having great value

8. severe — very harsh or difficult; stern

9. century — a period of 100 years

10. barren — not producing anything; bare

COMPLETE THE SENTENCES

>>>> *Use the vocabulary words in this lesson to complete the following sentences. Use each word only once.*

precious	penal	bleak	rarely	economy
severe	exile	century	colony	barren

1. When people see the _____bleak_____ landscape of Siberia, they usually wonder how anyone survives there.

2. Land that is _____barren_____ is not usually good for farming.

3. Being sent to a prison _____colony_____ was punishment for speaking out against the government.

4. People who were sent to prison colonies _____rarely_____ returned to their homes.

5. Settlers were unaware that _____precious_____ metals lay beneath their feet.

6. It would have helped the Soviet _____economy_____ to make better use of the land.

7. After a _____severe_____ winter, even a short summer must feel good to the people in Siberia.

8. Siberia is thought of as a place of _____exile_____ rather than a land of promise.

9. There are very sad stories written about life in a _____penal_____ colony.

10. Over the last _____century_____, important changes have taken place in Siberia.

USE YOUR OWN WORDS

>>>> *Look at the picture. What words come into your mind other than the ten vocabulary words used in this lesson? Write them on the lines below. To help you get started, here are two good words:*

1. _____children_____
2. _____coats_____
3. ___Answers will vary.___
4. _____
5. _____
6. _____
7. _____
8. _____
9. _____
10. _____

>>>> **Look at the vocabulary words penal and economy. Together, they are made up of 12 letters. See how many words you can form by using the letters of these words. Make up at least ten words. Write your words in the spaces below.**

penal

1. _Answers will vary._

2. _____

3. _____

4. _____

5. _____

economy

6. _____

7. _____

8. _____

9. _____

10. _____

COMPLETE THE STORY

>>>> Here are the ten vocabulary words for this lesson:

bleak	rarely	precious	economy	century
colony	exile	penal	severe	barren

>>>> **There are five blank spaces in the story below. Five vocabulary words have already been used in the story. They are underlined. Use the other five words to fill in the blanks.**

In the <u>century</u> before last, Russia used the area of Siberia as a _____penal_____ <u>colony</u> for criminals and political prisoners. The inhabitants of the land were mainly native Asians, Russian explorers, and prisoners. For the prisoners, it was bad enough to be forced into _____exile_____, but cold and hunger made it worse. The <u>severe</u> winters killed many of them.

Siberia is a land rich in oil, coal, and lumber, as well as silver, gold, and other <u>precious</u> metals. The government knew that this wealth was necessary for the nation's _____economy_____. But there were _____rarely_____ enough Siberians to bring it out. So the government tried to encourage other citizens to go live and work in the _____bleak_____ and <u>barren</u> land.

Learn More About Siberia

>>>> *On a separate piece of paper or in your notebook or journal, complete one or more of the activities below.*

Learning Across the Curriculum

Research the geography of Siberia. Write a description of the area for someone who has never heard of it and wants to travel there. Include the different kinds of land features in Siberia, its boundaries, and what the weather is like at different times of year. Share your description with a classmate.

Broadening Your Understanding

The Trans-Siberian Railway is more than a hundred years old. Today, travelers still take the train through Siberia on this route. Imagine that you are planning a trip through Siberia on the railway. Look in travel and write a summary of your trip. What will you be seeing? What cities and towns will you go through? What will the highlights of the trip be?

Appreciating Diversity

Find out about the different ethnic groups that live in Siberia, such as the Tungus, Evenkis, Okrugs, and Yakuts. How have these people adapted to living in such a harsh climate? Describe how they live. Give an oral presentation to the class.

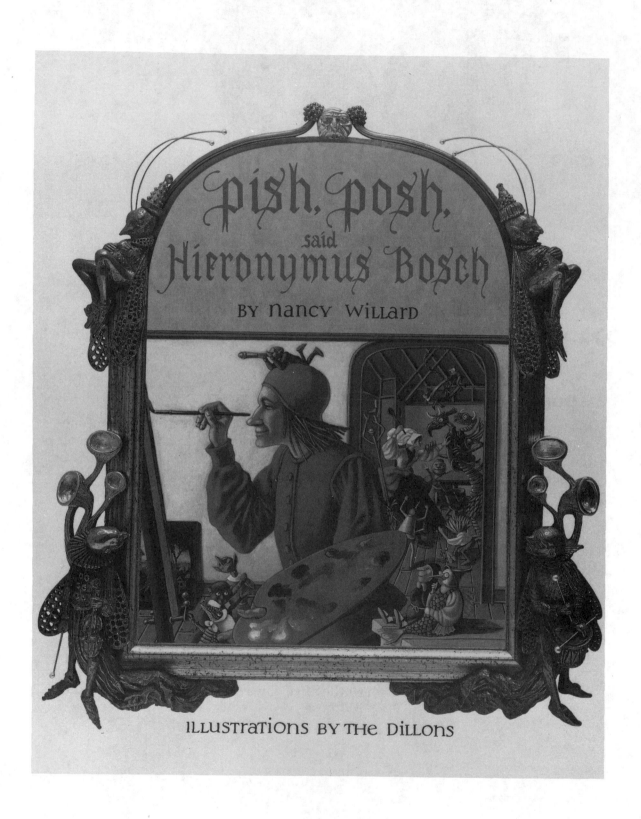

pish, posh, said Hieronymus Bosch

By Nancy Willard

ILLUSTRATIONS BY THE DILLONS

15 A DOUBLE VISION

Leo and Diane Dillon are talented artists. They use their skills to illustrate children's books. As illustrators, the Dillons design pictures to accompany the words of a story. These detailed pictures make the story come alive.

The Dillons have worked together for more than 35 years. They have a system for illustrating a book. The first step is reading the text of the story. Then they discuss ideas for pictures. Once they agree on an idea, they begin to sketch it out. They take turns working on the drawing. They blend their skills to create the sketch. The result is a combination of the two artists' styles.

The pictures produced by this system are interesting. Generally, the characters' faces seem quite real. This is because the Dillons often use real people as models for their characters. Sometimes, the artists draw themselves in their pictures. They have even included their pet cats in their sketches.

Through their hard work, the Dillons have gained the admiration of many writers. They are constantly asked to work on new books. However, this success did not come easily. The Dillons admit that their career has been a long, slow struggle. Yet, they never gave up. They continued to cooperate on ideas as a team. After 35 years of teamwork, they are enjoying the success they deserve.

UNDERSTANDING THE STORY

>>>> *Circle the letter next to each correct statement.*

1. The story shows that
   a. it is difficult to work with a partner.
   b. all artists should work in pairs.
   c. teamwork pays off.

2. From this story, you can conclude that
   a. the Dillons want to write a book.
   b. the Dillons enjoy their work.
   c. the Dillons are allergic to dogs.

MAKE AN ALPHABETICAL LIST

>>>> *Here are the ten vocabulary words in the lesson. Write them in alphabetical order in the spaces below.*

system	sketch	constantly	detailed	generally
accompany	cooperate	discuss	admiration	included

1. accompany
2. admiration
3. constantly
4. cooperate
5. detailed

6. discuss
7. generally
8. included
9. sketch
10. system

WHAT DO THE WORDS MEAN?

>>>> *Following are some meanings, or definitions, for the ten vocabulary words in this lesson. Write the words next to their definitions.*

1. admiration ___ a feeling of wonder and approval

2. sketch ___ incomplete drawing

3. detailed ___ having many small parts

4. accompany ___ to go along with

5. discuss ___ to talk over

6. system ___ a plan; a set of rules

7. cooperate ___ to work together

8. generally ___ usually; ordinarily

9. included ___ made part of; involved

10. constantly ___ repeatedly; happening again and again

88

COMPLETE THE SENTENCES

>>>> *Use the vocabulary words in this lesson to complete the following sentences. Use each word only once.*

generally	detailed	admiration	cooperate	constantly
accompany	sketch	included	system	discuss

1. Leo and Diane Dillon have a _____ system _____ for illustrating a book.

2. They _____ cooperate _____ on all the work.

3. After reading the story, they _____ discuss _____ ideas for pictures.

4. They take turns working on a _____ sketch _____.

5. Their illustrations are quite _____ detailed _____.

6. They have _____ included _____ sketches of their cats in their illustrations.

7. The Dillons _____ generally _____ use real people as models for their work.

8. Many authors want the Dillons to draw pictures to _____ accompany _____ their stories.

9. The Dillons are _____ constantly _____ asked to illustrate new books.

10. Their hard work won them the _____ admiration _____ of many fans.

USE YOUR OWN WORDS

>>>> *Look at the picture. What words come into your mind other than the ten vocabulary words used in this lesson? Write them on the lines below. To help you get started, here are two good words:*

1. _____ creative _____

2. _____ illustration _____

3. _____ Answers will vary. _____

4. _____

5. _____

6. _____

7. _____

8. _____

9. _____

10. _____

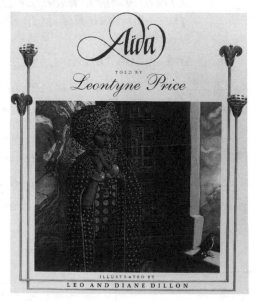

CIRCLE THE SYNONYMS

▸▸▸▸ Do you remember what a **synonym** is? It is a word that means the same or nearly the same as another word. *Unhappy* and *sad* are synonyms.

▸▸▸▸ *Six of the vocabulary words in this lesson are listed below. To the right of each vocabulary word are three other words or groups of words. Two of them are synonyms for the vocabulary word. Draw a circle around the two synonyms for each vocabulary word.*

Vocabulary Words   **Synonyms**

1. **cooperate**   disagree   (work together) (unite)
2. **system**   (plan)   machine   (method)
3. **generally**   (usually)   (often)   never
4. **constantly**   (repeatedly) rarely   (often)
5. **discuss**   (talk)   be silent   (speak about)
6. **admiration**   hatred   (wonder)   (respect)

COMPLETE THE STORY

▸▸▸▸ Here are the ten vocabulary words for this lesson:

accompany	admiration	detailed	generally	system
cooperate	discuss	included	sketch	constantly

▸▸▸▸ *There are five blanks in the story below. Five vocabulary words have already been used in the story. They are underlined. Use the other five words to fill in the blanks.*

Leo and Diane Dillon _____cooperate_____ to illustrate children's books. They _____generally_____ follow a <u>system</u> when working on a new book. They read the book. They come up with ideas for pictures to <u>accompany</u> the story. They _____discuss_____ their ideas. They take turns working on each _____sketch_____. The <u>detailed</u> drawings produced by this <u>system</u> are quite unique. Through their sketches, the Dillons have gained the _____admiration_____ of the publishing world. Many authors want the Dillons' drawings <u>included</u> in their books.

Learn More About The Dillons

>>>> *On a separate piece of paper or in your notebook or journal, complete one or more of the activities below.*

Broadening Your Understanding

In this lesson, you discovered that Leo and Diane Dillon often use real people as models for the characters they draw. Suppose you needed to illustrate a book about a person your own age. Who would you use as a real-life model for this character? Write a brief paragraph identifying whom you would select as your model and why you chose this person.

Learning Across the Curriculum

The Dillons are quite famous for their illustrations of books based on African tales or traditions. Use reference texts to find a folktale from another culture. Share the folktale with your classmates. Make a poster that illustrates one part of the tale.

Extending Your Reading

In a library, find the books below or others illustrated by the Dillons. Look at the similarities and the differences of the Dillons' art styles used in the books. Read one of the books and write a paragraph telling how you think the Dillons' pictures help to tell the story.

Northern Lullaby, by Nancy White
The People Could Fly: American Black Folktales, by Virginia Hamilton
Many Thousand Gone, by Virginia Hamilton
Why Mosquitoes Buzz in People's Ears, by Verna Aardema
Aida, by Leontyne Price

Two people come to the center of the mat. They bow. This movement shows respect for each other and for their sport. They have practiced for hours. Now, they are ready. The judges are also ready. This contest is a show of skill. This sport is karate.

Karate is a Japanese word. It means "open hand." This is because it is never violent. It is *violent* only in the movies and on TV. What is shown on TV and in movies is not true karate.

The *origin* of this art *form* is *uncertain.* Many people believe it was started by a Buddhist monk in *remote* times. He taught it in China more than 1,300 years ago. Today, it has *spread* all over the world. It is taught in Japan. Many Koreans have become experts. People of the United States are studying it. Someday we may see karate included in the Olympics. Because of its true meaning, it may help people understand each other better.

Karate has three *purposes.* It develops a person's *spiritual* being. It *stimulates* the mind. It increases the body's *strength.* Karate is meant to build the mind, body, and spirit. It is used to protect life—never to destroy it.

UNDERSTANDING THE STORY

>>>> *Circle the letter next to each correct statement.*

1. The main idea of this story is that
   a. karate only exists on television or in the movies.
   b. a person must be religious to do karate.
   c. karate is a form of self-defense that has other benefits as well.

2. Though the story doesn't say so, karate, which means "open hand," probably gets its name from the fact that
   a. the object is to force the other person's open hand to the mat.
   b. hitting with an open hand is one of the main moves.
   c. a sword is held in the open hand.

MAKE AN ALPHABETICAL LIST

>>>> *Here are the ten vocabulary words in the lesson. Write them in alphabetical order in the spaces below.*

| form | purpose | strength | stimulates | violent |
| remote | origin | spiritual | spread | uncertain |

1. form
2. origin
3. purposes
4. remote
5. spiritual

6. spread
7. stimulates
8. strength
9. uncertain
10. violent

WHAT DO THE WORDS MEAN?

>>>> *Following are some meanings, or definitions, for the ten vocabulary words in this lesson. Write the words next to their definitions.*

1. origin — the beginning; where something comes from

2. uncertain — not known for sure; doubtful

3. remote — far off in time

4. spread — gone all over

5. form — a type; a kind

6. spiritual — holy or religious; having to do with the soul

7. strength — power; force

8. purposes — goals; aims

9. stimulates — excites; makes more active

10. violent — wild; roughly forceful

94

COMPLETE THE SENTENCES

>>>> Use the vocabulary words in this lesson to complete the following sentences. Use each word only once.

spread	violent	form	remote	spiritual
origin	uncertain	stimulates	strength	purposes

1. Karate is an ancient sport, surviving from _____remote_____ times.

2. The _____origin_____ of karate is not known, but it may have begun in China.

3. Karate _____stimulates_____ the mind and allows a person to think more clearly.

4. Karate has three _____purposes_____: to build the mind, body, and spirit.

5. The art of karate _____spread_____ from China to Japan and Korea.

6. Karate is considered a _____form_____ of art when it is done with grace and skill.

7. There is a _____spiritual_____, almost religious, quality to karate that people like.

8. We are _____uncertain_____ about the beginnings of karate, but we think it was started by a Buddhist monk.

9. Just knowing the _____strength_____ of your body has increased can make you feel safer.

10. Too often, the _____violent_____ side of karate is all that is shown on TV.

USE YOUR OWN WORDS

>>>> Look at the picture. What words come into your mind other than the ten vocabulary words used in this lesson? Write them on the lines below. To help you get started, here are two good words:

1. _____sash_____

2. _____fist_____

3. _____Answers will vary._____

4. _____

5. _____

6. _____

7. _____

8. _____

9. _____

10. _____

MATCH THE ANTONYMS

>>>> An **antonym** is a word that means the opposite of another word. *Fast* and *slow* are antonyms.

>>>> *Match the vocabulary words on the left with the antonyms on the right. Write the correct letter in the space next to the vocabulary word.*

Vocabulary Words			Antonyms
1.	sure	**uncertain**	a. gentle
2.	weakness	**strength**	b. sure
3.	gentle	**violent**	c. enclosed
4.	enclosed	**spread**	d. weakness
5.	near	**remote**	e. ending
6.	ending	**origin**	f. near

COMPLETE THE STORY

>>>> Here are the ten vocabulary words for this lesson:

origin	uncertain	spread	remote	form
strength	violent	spiritual	purposes	stimulates

>>>> *There are five blank spaces in the story below. Five vocabulary words have already been used in the story. They are underlined. Use the other five words to fill in the blanks.*

Although its _____origin_____ is <u>uncertain</u>, karate in one _____form_____ or another has <u>spread</u> across the United States.

Karate students are unsure about the _____purposes_____ of their art from <u>remote</u> times. They know it <u>stimulates</u> physical, mental, and _____spiritual_____ growth.

After a training session, most students promise: (1) to help each other develop spiritually, mentally, and physically; (2) to listen to all instruction; (3) to meet all problems with inner <u>strength</u>; (4) to be polite to all; (5) to remember the virtue of modesty; and (6) to use karate only to stop _____violent_____ acts.

Learn More About Martial Arts

>>>> *On a separate piece of paper or in your notebook or journal, complete one or more of the activities below.*

Learning Across the Curriculum

Martial arts, including karate, have a long history in Asian countries. Research the history of karate. Write a report about what you discover.

Broadening Your Understanding

Karate is only one of several kinds of martial arts, which also include aikido, judo, tae kwan do, and jujitsu. Find out what these different kinds of martial arts have in common and how they are different. Make a grid that compares the martial arts. The rows of the grid should list the name of the martial art. The columns of the grid should list history, country of origin, and how it is used today.

Extending Your Reading

Check out one of the following books on karate from the library. Each art has a philosophy behind it. Read one of the following books. Write about the philosophy that is taught, as well as the physical moves.

Karate Basics, by Allen Queen
Karate, by Jane Mersky Leder
Facing the Double-Edged Sword: The Art of Karate, by Doyle Terren Webster
Karate Handbook, by Allen Queen

17 LIFE ON A ROPE

Mountain climbing is tough on men and women. People who climb mountains must be well trained. They are mountaineers. The best trained become guides. A good leader and great teamwork are needed to prevent accidents or even death.

A special hammer is made just for mountain climbing. It has a hammerhead on one end and a pick on the other. The hammer is used to anchor a piton, or large spike, into the rock cliffs. This piton has a hole in one end. A rope can be threaded through this hole or a snap-ring can be attached. The snap-ring allows the rope to slide more easily. The pick end of the hammer is used most often to chip footholds and handholds in the cliffs or in ice fields, called glaciers.

The rope used in mountain climbing is made of nylon. It is used to raise and lower mountaineers up and down the mountain. Often their lives depend on this one thin nylon rope. These people must also be able to go over or around deep holes in the ice called crevasses.

The weather can often be very bad high up in the mountains. Sometimes there are blizzards. Great snowslides, called avalanches, may come tumbling down the mountainside. The climbers could freeze to death or be buried alive.

Even though they risk their lives on a rope, many people love this dangerous sport.

UNDERSTANDING THE STORY

>>>> *Circle the letter next to each correct statement.*

1. According to this story, in order to climb mountains a person must be
   a. at least 18 years old.
   b. able to run quickly.
   c. well trained.

2. When a sport is called dangerous, it means that
   a. there is a chance someone may be hurt or killed while doing it.
   b. it is especially difficult to do.
   c. people should not be allowed to do it.

MAKE AN ALPHABETICAL LIST

>>>> *Here are the ten vocabulary words in the lesson. Write them in alphabetical order in the spaces below.*

teamwork	mountaineers	anchor	guides	cliffs
snap-ring	snowslides	crevasses	blizzards	pick

1. _anchor_
2. _blizzards_
3. _cliffs_
4. _crevasses_
5. _guides_

6. _mountaineers_
7. _pick_
8. _snap-ring_
9. _snowslides_
10. _teamwork_

WHAT DO THE WORDS MEAN?

>>>> *Following are some meanings, or definitions, for the ten vocabulary words in this lesson. Write the words next to their definitions.*

1. _mountaineers_ people who climb mountains

2. _anchor_ to fix firmly; to hold fast

3. _snap-ring_ a metal ring with a clip for holding ropes

4. _snowslides_ masses of snow sliding down a mountainside; avalanches

5. _blizzards_ extreme and violent snowstorms

6. _teamwork_ working together; cooperation

7. _cliffs_ high mountain walls

8. _crevasses_ deep holes in rocks or ice

9. _guides_ leaders; persons who lead mountain climbers

10. _pick_ a pointed tool for making holes in rocks

COMPLETE THE SENTENCES

>>>> *Use the vocabulary words in this lesson to complete the following sentences. Use each word only once.*

anchor	crevasses	cliffs	mountaineers	teamwork
guides	pick	snowslides	snap-ring	blizzards

1. __Mountaineers__ are people who climb mountains for sport.

2. The guide warned us to watch out for __snowslides__ after it had snowed steadily for 24 hours.

3. I have the names of several __guides__ who can take us climbing and "show us the ropes."

4. __Blizzards__ are a combination of strong winds and heavy snows.

5. One guide showed us how to __anchor__ a large spike, or piton.

6. It took extra effort to climb the steep __cliffs__ that were covered with snow and ice.

7. No matter how strong each person is, a group needs __teamwork__ if it is going to make it to the top of a high mountain.

8. The mountaineer used a __pick__ to dig footholds in the cliff.

9. The __snap-ring__ was greased so that the rope could slide easily.

10. You must watch out for deep holes, or __crevasses__; they are dangerous.

USE YOUR OWN WORDS

>>>> *Look at the picture. What words come into your mind other than the ten vocabulary words used in this lesson? Write them on the lines below. To help you get started, here are two good words:*

1. __ropes__
2. __hands__
3. __Answers will vary.__
4. _____
5. _____
6. _____
7. _____
8. _____
9. _____
10. _____

CIRCLE THE SYNONYMS

>>>> Do you remember what a **synonym** is? It is a word that means the same or nearly the same as another word. *Sad* and *unhappy* are synonyms.

>>>> *Six of the vocabulary words for this lesson are listed below. To the right of each word are three words or phrases. Two of them are synonyms for the vocabulary word. Draw a circle around the two synonyms for each vocabulary word.*

Vocabulary Words	Synonyms		
1. **teamwork**	(cooperation)	(interaction)	selfishness
2. **crevasses**	(openings)	seams	(holes)
3. **anchor**	lean	(secure)	(fasten)
4. **guides**	(leaders)	followers	(teachers)
5. **blizzards**	(snowstorms)	(winter storms)	warm winds
6. **cliffs**	little hills	(mountainsides)	(rock walls)

COMPLETE THE STORY

>>>> Here are the ten vocabulary words for this lesson:

teamwork	mountaineers	anchor	guides	cliffs
snap-ring	snowslides	crevasses	blizzards	pick

>>>> *There are five blank spaces in the story below. Five vocabulary words have already been used in the story. They are underlined. Use the other five words to fill in the blanks.*

An old and exciting sport is mountain climbing. <u>Mountaineers</u> should never start without training and practice. They should listen carefully to their ___guides___. These leaders teach the climbers how to use the special tools, such as the <u>pick</u> end of the hammer. Climbers must learn how to ___anchor___ ropes to <u>cliffs</u> with a piton and a ___snap-ring___. They also learn how to cross or go around ___crevasses___. In winter climbing, they learn the dangers of <u>snowslides</u> and ___blizzards___. One of the most important lessons they learn is <u>teamwork</u>. Mountain climbing is a sport for groups, not individuals.

102

Learn More About Mountain Climbing

>>>> *On a separate piece of paper or in your notebook or journal, complete one or more of the activities below.*

Working Together

Work with a group to make a mural that explains the history of mountain climbing in North America. Draw a large map of North America. Divide up the continent. Have different people research the major mountains that have been climbed. Find out the height of each mountain, who climbed it, and when. Write this information near the place the mountain appears on the map.

Broadening Your Understanding

Describe a properly dressed mountain climber. Research all the equipment a mountain climber needs to climb a mountain safely. Write an equipment list, including the use of each piece of equipment. Or draw an illustration of a fully equipped mountain climber, with each piece of equipment identified and explained in the picture.

Learning Across the Curriculum

Mountain climbers face dangerous conditions. Research the medical problems mountain climbers most often face. Write what climbers can do to prevent these problems and what they can do if they experience them.

Did you ever wonder what a person does after being President? If you are Jimmy Carter, you work on a home. Not your home, but a home for a low- income family.

After leaving office, Carter was approached by Millard Fuller. Fuller asked Carter to lend a hand to a program he had started, called Habitat for Humanity. Fuller founded the program in 1976. The goal of Habitat for Humanity is to build homes for low-income families. The group purchases land and building supplies. Volunteers donate their time to build houses. The houses are sold at reduced prices. However, there is one catch. The buyers must give 500 hours of their own time to building new houses!

Carter consented to help the program. He and his wife Rosalynn spend one week each year working on houses. They work alongside other less-famous volunteers. They hammer nails and saw wood. They do whatever is needed to get the job done.

The hard work of these volunteers has certainly paid off. More than 20,000 homes have been built by this program. Construction of these new homes is not limited to the United States. Habitat for Humanity homes have been built in 40 different countries. Yet, there is still more work to be done. The program plans to continue until every person has a decent place to live.

UNDERSTANDING THE STORY

>>>> *Circle the letter next to each correct statement.*

1. Another good title for this story would be:
   a. "Presidents of the United States."
   b. "Putting a Roof over Those Without Homes."
   c. "Houses of America."

2. Though it doesn't say so, from the story you can tell that
   a. Jimmy Carter enjoys helping people.
   b. Millard Fuller hopes to become President.
   c. Jimmy Carter has found little to do since leaving the presidency.

>>>> Here are the ten vocabulary words in the lesson. Write them in alphabetical order in the spaces below.

| income | founded | approached | consented | purchases |
| decent | volunteers | goal | reduced | lend |

1. approached
2. consented
3. decent
4. founded
5. goal

6. income
7. lend
8. purchases
9. reduced
10. volunteers

WHAT DO THE WORDS MEAN?

>>>> Following are some meanings, or definitions, for the ten vocabulary words in this lesson. Write the words next to their definitions.

1. consented — agreed
2. purchases — buys
3. decent — good enough; suitable
4. reduced — lowered; diminished
5. income — money received for work
6. goal — the purpose; the objective
7. volunteers — people who give aid and services for free
8. founded — created; set up
9. lend — to give; to provide
10. approached — reached; contacted

COMPLETE THE SENTENCES

>>>> *Use the vocabulary words in this lesson to complete the following sentences. Use each word only once.*

| income | reduced | purchases | volunteers | consented |
| approached | lend | goal | founded | decent |

1. Habitat for Humanity was _____founded_____ by Millard Fuller.

2. The main _____goal_____ of the program is to provide affordable housing for all people.

3. Houses built through the program are sold to low-_____income_____ families.

4. The construction materials are _____purchased_____ by the program.

5. The homes are sold at _____reduced_____ prices.

6. Many _____volunteers_____ donate their time to help the program.

7. Carter and his wife _____lend_____ a hand to the program.

8. They had been _____approached_____ by Millard Fuller.

9. People who buy the homes have _____consented_____ to spend 500 hours building other homes.

10. Through the program, many families now have _____decent_____ places to live.

USE YOUR OWN WORDS

>>>> *Look at the picture. What words come into your mind other than the ten vocabulary words used in this lesson? Write them on the lines below. To help you get started, here are two good words:*

1. _____construction_____
2. _____President_____
3. _____Answers will vary._____
4. _____
5. _____
6. _____
7. _____
8. _____
9. _____
10. _____

IDENTIFY THE ANTONYMS AND SYNONYMS

▶▶▶▶ There are six vocabulary words listed below. To the right of each is either a synonym or an antonym. Remember: a **synonym** is a word that means the same or nearly the same as another word. An **antonym** is a word that means the opposite of another word.

▶▶▶▶ *On the line beside each pair of words, write S for synonyms or A for antonyms.*

Vocabulary Words	Antonyms and Synonyms	
1. **founded**	destroyed	1. _A_
2. **goal**	purpose	2. _S_
3. **lend**	give	3. _S_
4. **reduced**	increased	4. _A_
5. **consented**	refused	5. _A_
6. **decent**	suitable	6. _S_

COMPLETE THE STORY

▶▶▶▶ Here are the ten vocabulary words for this lesson:

decent	income	consented	purchases	reduced
founded	volunteers	approached	goal	lend

▶▶▶▶ *There are five blanks in the story below. Five vocabulary words have already been used in the story. They are underlined. Use the other five words to fill in the blanks.*

In 1976, Millard Fuller _____founded_____ Habitat for Humanity. The <u>goal</u> of this program is to provide _____decent_____, affordable housing for all people. Habitat for Humanity _____purchases_____ land and building materials at <u>reduced</u> prices. _____Volunteers_____ use the materials to build houses. The houses are sold to low-<u>income</u> families.

Participation in the program is not limited to everyday citizens. Some of the volunteers are quite famous. Former President Jimmy Carter and his wife Rosalynn _____lend_____ a hand. They became involved when Fuller <u>approached</u> Mr. Carter. The Carters <u>consented</u> to help out. One week each year, the former President swings a hammer alongside other volunteers.

108

Learn More About Building Homes

>>>> *On a separate piece of paper or in your notebook or journal, complete one or more of the activities below.*

Learning Across the Curriculum

Imagine you have been asked to design the next house that Habitat for Humanity will build. Draw a plan that shows the layout of the house. Share your plan with the class. Explain why you decided upon the particular plan.

Broadening Your Understanding

Before starting Habitat for Humanity, Millard Fuller was a successful lawyer. Do research to find out what caused Fuller to start the program. Share your findings in an oral report to the class.

Learning Across the Curriculum

Do research to learn more about Jimmy Carter's term in office. Identify major events that occurred during the years he was President. Create a time line that highlights the accomplishments of the 39th President.

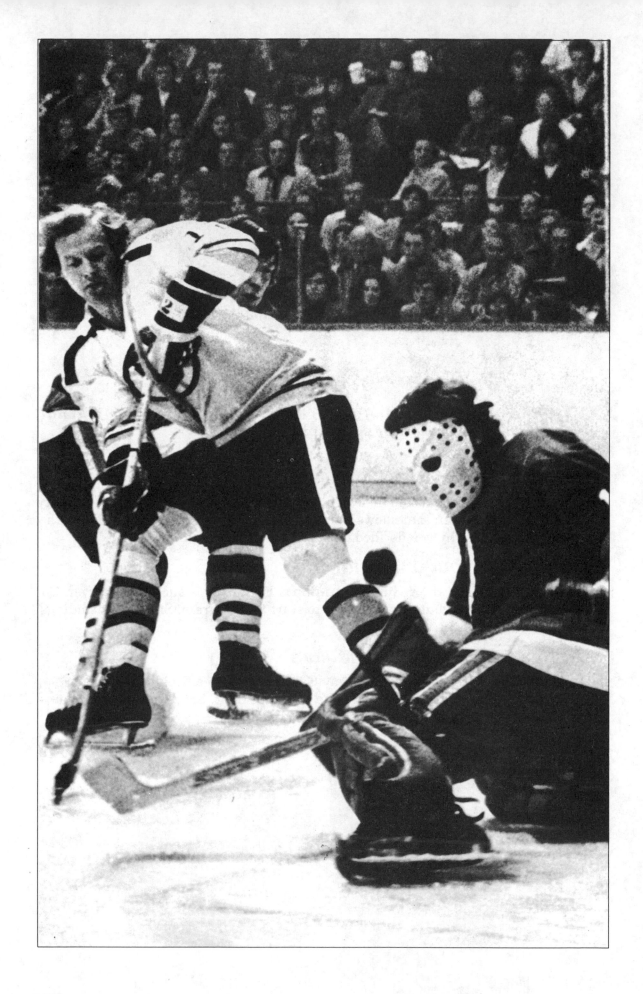

The player in the light jersey has just stolen the **puck!** He **zooms** in on the **goalie.** The other players have not gotten down the ice yet. It is player against player and goalie **versus** shooter. Will the player score for his team, or will the goalie **block** the shot?

Hockey is a **rough** sport. A player can easily be cut by a stick or knocked down on the ice. The puck is made of hard rubber. It is frozen before each game. This process makes the puck slide better on the ice. It also makes it hard as a rock. When shot at the net, the puck travels more than 100 miles an hour. In a single game, a goalie may have to block as many as 30 or 40 shots. This action takes great skill. It also takes great **courage.** Even with pads on, stopping a puck going that fast can hurt. It is no wonder that it is usually the goalie who has the most **bruises.** Some goalies even **suffer** more serious injuries.

What makes a person want to be a goalie? Most goalies will tell you that they like being where the action is. They like knowing that the whole team **depends** on them. They like doing a tough job well.

Look at that! He blocked the shot! Now he can rest a minute. Soon the players will be down at this end of the ice again, and the excitement will start all over.

UNDERSTANDING THE STORY

>>>> *Circle the letter next to each correct statement.*

1. The main idea of this story is that
   a. hockey is a rough but exciting sport, especially for goalies.
   b. hockey pucks travel more than 100 miles per hour.
   c. hockey is played one-to-one rather than as a team.

2. Though this story doesn't say so, goalies probably get hurt less often these days because
   a. better pads and masks are available.
   b. hockey sticks are no longer allowed in the game.
   c. new cures for cuts and bruises have been discovered.

MAKE AN ALPHABETICAL LIST

>>>> *Here are the ten vocabulary words in the lesson. Write them in alphabetical order in the spaces below.*

| rough | bruises | suffer | zooms | depends |
| versus | goalie | block | puck | courage |

1. block
2. bruises
3. courage
4. depends
5. goalie

6. puck
7. rough
8. suffer
9. versus
10. zooms

WHAT DO THE WORDS MEAN?

>>>> *Following are some meanings, or definitions, for the ten vocabulary words in this lesson. Write the words next to their definitions.*

1. suffer — to experience something painful; to put up with

2. rough — harsh; difficult; somewhat violent

3. block — to stop something

4. goalie — a person who guards the goal in hockey

5. puck — a hard, rubber disk used in ice hockey

6. bruises — marks caused by an injury that does not break the skin

7. depends — needs; relies on

8. courage — an ability to face danger or difficulty; bravery

9. versus — against

10. zooms — moves quickly

COMPLETE THE SENTENCES

>>>> **Use the vocabulary words in this lesson to complete the following sentences. Use each word only once.**

rough	bruises	suffer	zooms	depends
versus	goalie	block	puck	courage

1. The goalie was able to _____ block _____ the first shot.

2. It was our goalie _____ versus _____ their star player.

3. The _____ puck _____ was moving very fast across the ice.

4. If you don't wear a mask when you play hockey, you may _____ suffer _____ a head or face injury.

5. The whole team _____ depends _____ on the goalie to stop net shots.

6. Believe me, it takes a lot of _____ courage _____ to stand in front of the net with sticks flying all around.

7. The _____ bruises _____ on the goalie's leg looked bad, but they weren't serious.

8. Do you see how that player _____ zooms _____ down the ice behind the puck?

9. Hockey can be a very _____ rough _____ sport, so players always wear padding.

10. To be a _____ goalie _____, a person has to have quick reflexes and courage.

USE YOUR OWN WORDS

>>>> **Look at the picture. What words come into your mind other than the ten vocabulary words used in this lesson? Write them on the lines below. To help you get started, here are two good words:**

1. _____ ice _____
2. _____ players _____
3. _____ Answers will vary. _____
4. _____
5. _____
6. _____
7. _____
8. _____
9. _____
10. _____

UNSCRAMBLE THE LETTERS

>>>> Each group of letters contains the letters in one of the vocabulary words for this lesson. Can you unscramble them? Write your answers in the lines to the right of each letter group.

Scrambled Words	Vocabulary Words
1. clbok	block
2. ruegcoa	courage
3. sdeepnd	depends
4. oghur	rough
5. mosoz	zooms
6. sisrube	bruises
7. ukpc	puck
8. lageoi	goalie
9. esurvs	versus
10. feurfs	suffer

COMPLETE THE STORY

>>>> Here are the ten vocabulary words for this lesson:

courage	zooms	goalie	suffer	rough
versus	block	puck	depends	bruises

>>>> There are five blank spaces in the story below. Five vocabulary words have already been used in the story. They are underlined. Use the other five words to fill in the blanks.

Hockey is an action-packed sport. It can also be <u>rough</u>. Often, players _____suffer_____ injuries. Most of these injuries are just _____bruises_____, but some are more serious. The player who probably has it the roughest is the _____goalie_____. Goalies are expected to <u>block</u> shots from going into the net at all costs. The <u>puck</u> is hard and travels at a very high speed. Usually, goalies catch the puck in their huge gloves. But sometimes they must throw themselves down on the ice to block a shot. This action takes _____courage_____! Most of the time, at least one other player <u>zooms</u> down the ice to help the goalie defend the goal. At times, however, it seems to be the goalie _____versus_____ the other team. But most goalies enjoy the challenge. They like the excitement. They also like knowing that each of their teammates <u>depends</u> on them.

Learn More About Hockey

>>>> *On a separate piece of paper or in your notebook or journal, complete one or more of the activities below.*

Building Language

Look for an article from the newspaper sports section about a hockey game. Underline unfamiliar phrases in the article. Write what you think the phrase means. Check your work with another student or a dictionary.

Broadening Your Understanding

The game of hockey can be rough, and players—particularly goalies—wear equipment to protect themselves. Find out more about a goalie's equipment by interviewing a clerk at a sporting goods store. Find out what equipment is necessary, how it is made, and how it helps to protect a goalie. Write a report about what you discover.

Extending Your Reading

Wayne Gretzky is one of the most famous hockey players. Read one of the following books about him. What helped him succeed? What are some highlights of his career? Write your answers as if you were writing a paragraph in a program for a hockey game.

Gretzky, Gretzky, Gretzky!, by Meguido Zola
Wayne Gretzky, by Bert Rosenthal
Wayne Gretzky, by Thomas R. Raber

20 THE BULLFIGHT

Ay! Torero! It is Sunday afternoon in Spain. A brave man faces a brave bull. The bull is huge, much larger than the man. Its horns are sharp. The $\boxed{\textit{matador}}$ has fought many bulls before. He is well-trained and experienced. Man and bull both have $\boxed{\textit{advantages.}}$ The bull's feet move quickly. It $\boxed{\textit{charges}}$ the matador. The matador spins $\boxed{\textit{gracefully}}$ one way. He waves his $\boxed{\textit{cape}}$ another. The bull follows the cape. The watching fans cheer.

Then the matador's coworkers push sharp $\boxed{\textit{spears}}$ into the bull's neck and shoulders. Its head drops low. The bull charges again. But the matador moves on his toes like a $\boxed{\textit{ballet}}$ dancer. The bull misses—it does not $\boxed{\textit{gore}}$ him. The horns come very close, and the crowd shouts, *"Ole!"* The bull tries again and again. The matador moves quickly out of his way each time.

The bull was $\boxed{\textit{bred}}$ to be strong and brave. Its father was strong and brave. So was its father's father. It is not afraid of anything. The bull charges once more. The matador raises his sword. He thrusts it into the neck of the bull, behind its head. The bull is dead. Another day, the matador might die.

One out of every three great matadors has died a $\boxed{\textit{cruel}}$ death on the horns of a brave bull.

UNDERSTANDING THE STORY

 Circle the letter next to each correct statement.

1. Another good title for this story might be:
   a. "The Mighty Matador."
   b. "Ballet in the Sun."
   c. "A Sunday Afternoon."

2. Though it doesn't say so, from the story you can tell that large crowds of people watch the bullfights because
   a. no other sports are permitted on Sunday.
   b. the contest between matador and bull is very exciting.
   c. friends of the matadors get free passes.

MAKE AN ALPHABETICAL LIST

>>>> *Here are the ten vocabulary words in the lesson. Write them in alphabetical order in the spaces below.*

matador	bred	gracefully	charges	gore
spears	advantages	cruel	cape	ballet

1. advantages
2. ballet
3. bred
4. cape
5. charges

6. cruel
7. gore
8. gracefully
9. matador
10. spears

WHAT DO THE WORDS MEAN?

>>>> *Following are some meanings, or definitions, for the ten vocabulary words in this lesson. Write the words next to their definitions.*

1. matador _____ a person who kills the bull in a bullfight

2. advantages _____ things that help you; benefits

3. spears _____ weapons that are long poles with sharp metal points

4. bred _____ produced and raised with special qualities

5. gracefully _____ beautifully and smoothly

6. cape _____ a cloth worn over the shoulders; used in bullfighting to fool the bull

7. ballet _____ a graceful dance

8. gore _____ to stab with a horn; to injure

9. charges _____ attacks; runs at something with speed

10. cruel _____ not kind; brutal

118

F

famous *[FAY muhs]* well known by many people
fans *[FANZ]* people who are enthusiastic about a performer
fashion *[FASH uhn]* current style of dress
females *[FEE maylz]* girls; women
field *[FEELD]* the type of job one does
fins *[FINZ]* rubber flippers that people wear on their feet to swim and dive
foreign *[FOR in]* coming from another country
form *[FORM]* a type; a kind
founded *[FOUN duhd]* created; set up

G

generally *[JEN uhr uhl ee]* usually; ordinarily
glow *[GLOH]* a light; to shine
goal *[GOHL]* the purpose; the objective
goalie *[GOH lee]* a person who guards the goal in hockey
gondola *[GON duh luh]* a car or basket hung under a balloon
gondolas *[GON duh luz]* long, narrow boats with a high peak at each end
gore *[GOR]* to stab with a horn; to injure
gracefully *[GRAYS fuh lee]* beautifully and smoothly
guest *[GEST]* visitor
guides *[GYDZ]* leaders; persons who lead mountain climbers
gusts *[GUSTS]* violent rushes; sudden outbursts

H

habit *[HAB it]* an outfit worn by a nun
helicopter *[HEL i kawp tuhr]* aircraft with circular blades
helium *[HEE lee um]* gas used to fill balloons
hired *[HYRD]* given a job
honest *[ON est]* truthful; not phony
host *[HOHST]* the main announcer on a show; someone giving a party for invited guests
hush *[HUSH]* quiet, sudden silence

I

imaginary *[i MAJ i ner ee]* existing only in the mind or imagination; unreal
improvise *[IM pruh veyez]* to make up
included *[in KLOO uhd]* made part of; involved
income *[IN kuhm]* money received for work
inflated *[in FLAYT id]* filled up
inhabitants *[in HAB uh tunts]* persons or animals who live in a place
interested *[IN trist id]* wanting to know more about something; concerned
ironic *[eye RON ik]* something that is opposite to what you would expect
islands *[I lands]* bodies of land surrounded by water

J

javelin *[JAV lin]* a spear used in sports events

K

key *[KEE]* most important; central

L

lend *[LEND]* to give; to provide

M

matador *[MAH tuh dor]* a person who kills the bull in a bullfight
mountaineers *[moun tuh NEERZ]* people who climb mountains
mouthpiece *[MOUTH pees]* a part of the scuba equipment that the diver holds in the mouth
movement *[MOOV munt]* crusade; organized effort
mumble *[MUM buhl]* to speak in an unclear way with the lips not open enough

N

national *[NASH uh nuhl]* having to do with a whole country
nature *[NAT chuhr]* the natural world
nervous *[NUR vus]* uneasy, uncomfortable
neurosurgery *[noor oh SUR jur ee]* operations on the brain, spinal cord, or nerves

O

observers *[uhb ZERV uhrs]* viewers
obstacles *[OB stih kulz]* barriers; difficulties
obtain *[uhb TAYN]* to gain possession of
origin *[OR uh jin]* the beginning; where something comes from

P

passengers *[PAS un jurz]* people who travel in a bus, boat, train, or plane
penal *[PEE nul]* involving punishment
pick *[PIK]* a pointed tool for making holes in rocks
pleasure *[PLEZH uhr]* delight; enjoyment
poverty *[PAWV uhr tee]* state of being poor
precious *[PRESH us]* having great value
prefer *[pree FUR]* to want one thing instead of another
preparation *[prep uh RAY shuhn]* readiness
previously *[PRE vee us lee]* earlier; before
pride *[PREYED]* good feelings about yourself; self-respect
professional *[proh FESH uhn uhl]* having to do with earning a living in a job that requires certain skills
provide *[pruh VEYED]* to give; to supply
puck *[PUK]* a hard, rubber disk used in ice hockey
purchases *[PER chuh sez]* buys
purposes *[PUR puh siz]* goals; aims

Q

qualify *[KWAHL uh feye]* to prove oneself worthy
quotas *[KWOH tuhs]* shares; positions held for a certain group

R

racism *[RAYS iz uhm]* a belief that one's own race is superior to another
rare *[RAIR]* scarce; only a few left
rarely *[RAIR lee]* not often; seldom
rate *[rayt]* measured quantity
reduced *[ri DOOSD]* lowered; diminished
regulated *[REG yuh layt id]* kept in working order; controlled
remote *[ree MOHT]* far off in time
respected *[rih SPEKT id]* admired
ridiculous *[ri DIK yuh luhs]* funny, silly
risks *[RISKS]* hazards; possibilities of danger
roam *[ROHM]* move about as one pleases; wander
rough *[RUF]* harsh; difficult; somewhat violent
rupture *[RUP chuhr]* break open; to burst

S

scheduled *[SKE joold]* happening at definite times
schools *[SKOOLZ]* large numbers of fish swimming together
screen *[SKREEN]* a surface or area on which movies or television images are shown
scuba *[SKOO buh]* gear that allows breathing underwater: *s*elf-contained *u*nderwater *b*reathing *a*pparatus
senior *[SEEN yur]* older; more than age 55
severe *[suh VEER]* very harsh or difficult; stern
shadows *[SHAD ohz]* areas of darkness or shade
sketch *[SKECH]* an incomplete drawing
skills *[SKILZ]* abilities that come from practice
smothered *[SMUTH uhrd]* to cut off oxygen supply; suffocated
snap-ring *[SNAP ring]* a metal ring with a clip for holding ropes
snowslides *[SNOH sleyedz]* masses of snow sliding down a mountainside; avalanches
sought *[SAWT]* searched for; tried to find
spears *[SPIRS]* weapons that are long poles with sharp metal points
species *[SPEE sheez]* animals that have some common characteristics or qualities
spiritual *[SPIR ih choo uhl]* holy or religious; having to do with the soul
spread *[SPRED]* gone all over
stimulates *[STIM yuh layts]* excites; makes more active
strength *[STRENTH]* power; force
stride *[STREYED]* a step or style of walking
style *[STEYEL]* a way of doing things
suffer *[SUF uhr]* to experience something painful; to put up with
surfacing *[SUR fuh sing]* coming to the top of the water
survival *[sur VEYE vul]* staying alive; existing
system *[SIS tuhm]* a plan; a set of rules

T

tasks *[TASKS]* jobs; assignments
teamwork *[TEEM wurk]* working together; cooperation

threat *[THRET]* something dangerous that might happen
tickets *[TIK uhts]* notices you get from a police officer for breaking the law

U

ultimately *[UHL tuh mit lee]* finally; at last
uncertain *[un SUR tuhn]* not known for sure; doubtful

V

various *[VER ee uhs]* different kinds; more than one
versus *[VUR suhs]* against
violent *[VEYE uh lunt]* wild; roughly forceful
visitors *[VIZ it uhrz]* people who visit; sightseers
volunteers *[vawl uhn TERZ]* people who give aid and services for free

W

water-bus *[WAW tur bus]* a canal boat with a motor that carries many passengers

Y

youngster *[yung stuhr]* child

Z

zooms *[ZOOMZ]* moves quickly